BEI GRIN MACHT SICH IHR WISSEN BEZAHLT

- Wir veröffentlichen Ihre Hausarbeit, Bachelor- und Masterarbeit

- Ihr eigenes eBook und Buch - weltweit in allen wichtigen Shops

- Verdienen Sie an jedem Verkauf

Jetzt bei www.GRIN.com hochladen und kostenlos publizieren

Bibliographic information published by the German National Library:

The German National Library lists this publication in the National Bibliography; detailed bibliographic data are available on the Internet at http://dnb.dnb.de .

Imprint:

Copyright © 2015 GRIN Verlag
Print and binding: Books on Demand GmbH, Norderstedt Germany
ISBN: 9783668965195

This book at GRIN:

https://www.grin.com/document/475233

Sadik Mejid

Synthesis of Chemical Compounds. Results, Discussion and Experimental Section

GRIN Verlag

Advanced Organic Chemistry

Experimental Module

Synthesis of Chemical compounds

Sadik Mejid

Department of Chemistry/Organic Chemistry

Faculty of Mathematics and Natural Sciences

University of Cologne

Cologne, 2015

Table of Contents

1 Results and discussion

1.1 Synthesis of *N*-(benzyloxycarbonyl)-(*S*)-proline (3)[1]

A literature survey *via Reaxys* revealed 2 references in which this compound was described as the product. The first synthesis has been published by Corey in the Journal of Organic Chemistry in 1988: *N*-(benzyloxycarbonyl)-(*S*)-proline **(3)** was synthesized in 75.2 % yield from proline by reaction with benzyl chloroformate in an aqueous solution at 0 – 5 °C. E. J. Corey, *J. Org. Chem.*, Vol. 53, No. 12, **1988**, 2861-2863 seemed to be the best choice for its high yield.

Scheme 1: reaction equation.

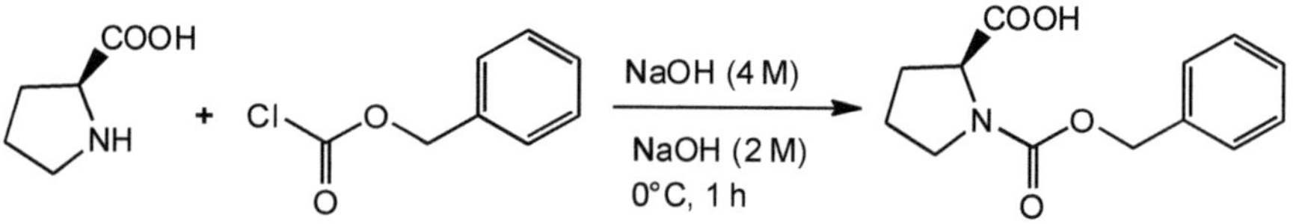

The protection of proline with benzylchloroformate gave a colourless solid, which recrystallized from petrolium ether. The yield of 75.2 % is below that reportedin the literature (96 %)[1]. The purity of the product could not be proofed, since no enough analytic data are available. For the product **(3)** no GC-MS was recorded, because of its carbon acid group that might damage the mass spectrometry system.

The hydroxide ion of NaOH deprotonates the NH-group of proline to form water and the electron pair of the nitrogen anion undergo a nucleophilic attack to the carbon atom of the carbonyl-group to give **(3)**. The sodium cation forms with chloride anion sodiumchloride that prcipitates (see scheme 2).

An example for the application of the cyclic amino acid is its use as a pharmaceutical intermidiate, which is used for the synthesis of Eletriptan, a drug for the treatment of migriane.[10]

Scheme 2: reaction mechanism:

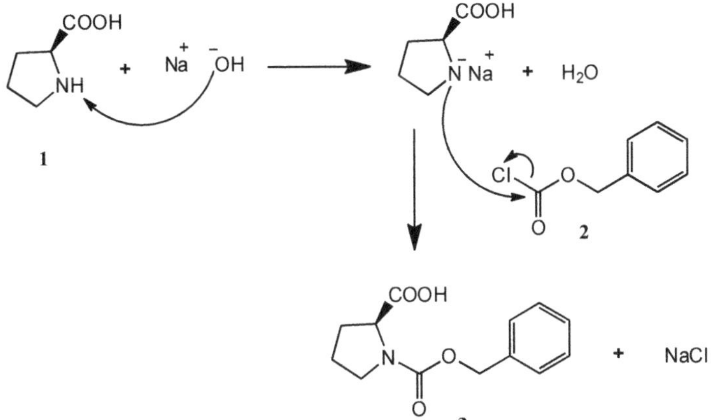

1.2 Synthesis of *N*-(benzyloxycarbonyl)-(*S*)-proline methyl ester (5)[1]

A literature survey *via Reaxys* revealed 2 references in which this compound was described as the product. The first synthesis has been published by Corey, in the Journal of Organic Chemistry in 1988: *N*-(benzyloxycarbonyl)-(*S*)-proline (3) was esterified in methanol with boron trifluoride etherate as catalyst to give the methyl ester (5) as an oil in 95 % yield. E. J. Corey, *J. Org. Chem.*, Vol. 53, No. 12, **1988**, 2861–2863 is the best choice for its short reaction time and also for its high yield, to the best of my knowledge.

Scheme 3: reaction equation.

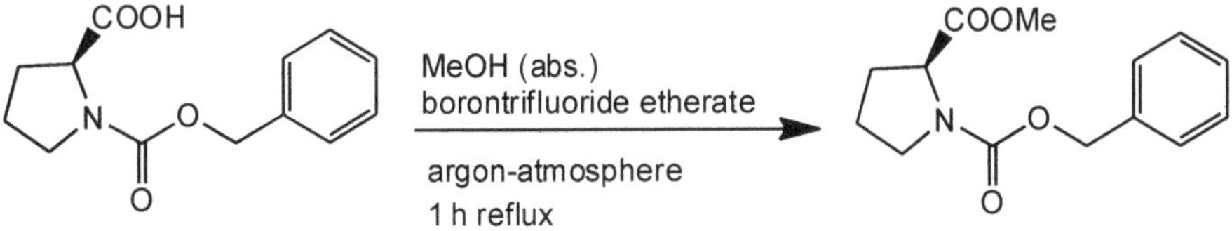

The esterification of *N*-(benzyloxycarbonyl)-(*S*)-proline with methanol has given *N*-(benzyloxycarbonyl)-(*S*)-proline methyl ester (5) as a yellowish oil in 95 % yield. The yield is much higher than the reported 85% yield of lit.[1]. The purity of the product is not satisfactory, since it is not a colourless oil as given in lit. The GC-MS shows the expected peak at t_R = 14.56 min. m/z 263.10 (4 %, [M]$^+$ calc.: 263,29), which proves the presence of the product.

The strong electron-withdrawing effect of the fluorine atoms on boron makes it highly electrophilic and lets it undergo nucleophilic attack by the electron pair of the oxygen of the carbonyl of protected proline. The carbon of the carbonyl group is now positively charged and will also undergoes nucleophilic attack by the oxygen of the methanol. After elimination of a water molecule and the boron trifluoride, *N*-(benzyloxycarbonyl)-(*S*)-proline methyl ester (**5**) is obtained as a colourless oil (see scheme 4).

Scheme 4: reaction mechanism.

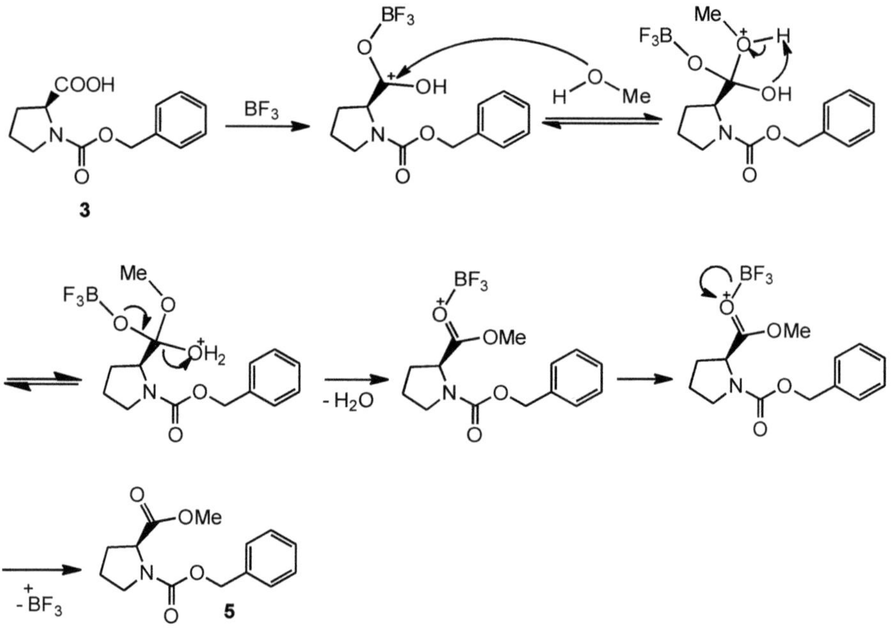

1.3 Synthesis of 2-amino-3-methylbutanol (valinol) (7)[3]

A literature survey *via Reaxys* revealed 12 references in which this compound was described as the product. The first synthesis has been published by Grotti, Moleculs, 2009: The amino acid L-Valine was reduced with lithium aluminum hydride in THF and 16 h under reflux at 70 °C to give 2-amino-3-methylbutanol (**7**) as yellow oil in 90 % yield. M. Grotli, *Molecules*, **2009**, 14, 5124 is the best choice due its high yield.

Scheme 1: reaction equation.

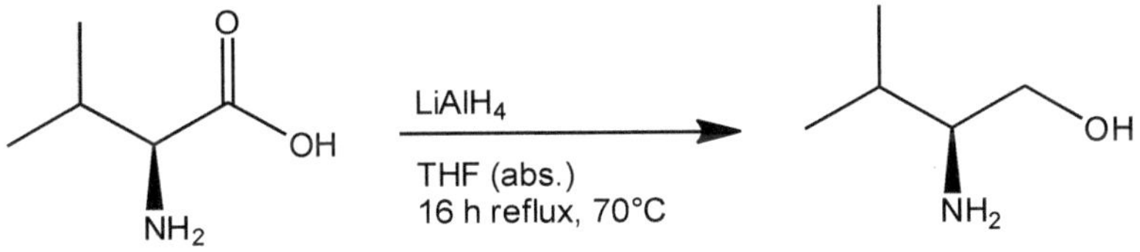

LiAlH$_4$

THF (abs.)
16 h reflux, 70°C

6

7

The reduction of L-Valine (6) with LiAlH$_4$ has given 2-amino-3-methylbutanol (7) as a yellow oil in 90 % yield. The yield is higher than that of litetartur of 80 %. The GC-MS shows the expected peak at t_R = 12.56 min. m/z 105.20 (22%, [M+2H]$^+$ calc.: 105.15), which can be assigned to the product, which proves its presence, Thus the available analytical data were consistent with the literature[4] and Thus the purity of the product is satisfactory.

The nucleophilic hydride of AlH$_4^-$ in the hydride reagent adds irreversible to the electrophilic C in the polar carbonyl group in the carboxylic acid, forming a metal alkoxide complex as an intermediate. Coordinative bonding of the carbonyl oxygen to a Lewis acidic metal (Li or Al) enhances that carbon's electrophilic character. This hydride addition is shown in scheme (5), with the hydride-donating part of the molecule being written as AlH$_4^-$. Although the lithium is not shown, it will be present in the products as a cationic component of ionic salts.[5]

Scheme 5: reaction mechanism.

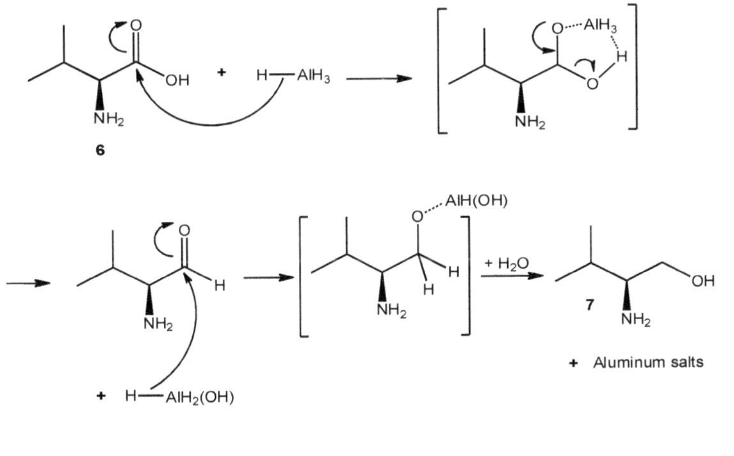

6

7

+ Aluminum salts

Valinol is mainly used to prepare chiral oxazolines, a process which can be achieved via a variety of methods. These oxazolines are principally used as ligands in asymmetric catalysis.[6]

1.4 Synthesis of (S)-4-isopropyl-2-oxazolidinone (10)[7]

A literature survey *via Reaxys* revealed 8 references in which this compound was described as the product. The first synthesis has been published by Benoit, Tetrahedron, Asymm., 2008: In an oxidative cyclocarbonylation Valinol (**7**) reacts with diethyl-carbonate (**8**) and after 3 h under reflux at 130 °C (S)-4-Isopropyl-2-oxazolidinone (**9**) as yellowish oil in 81 % yield was obtained. David Benoit, *Tetrahedron: Asymmetry* 19 (**2008**) 1068–1077 is the best choice to the best of my knowledge due to its high yield and the simplicity to carry out the reaction.

Scheme 6: reaction equation.

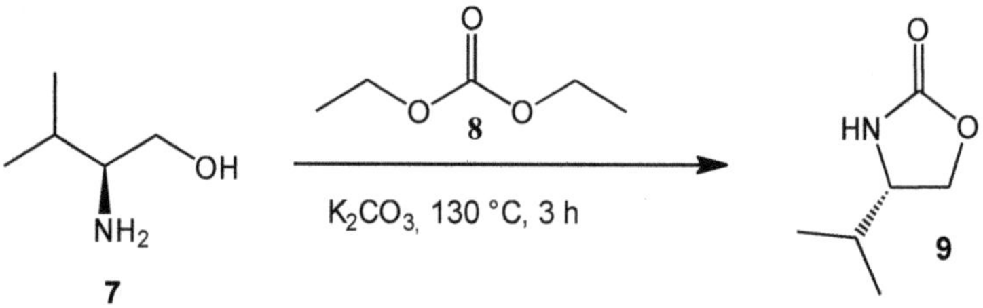

The oxidative cyclocarbonylation with diethylether (**8**) has given (S)-4-isopropyl-2-oxazolidinone (**9**) as a yellowish oil in 81 % yield. This yield is satisfactory, since the litetartur[7] promises 90 % as a colourless oil. The yellowish colour of the product could have been caused due to impurities, which could have been rid of through a destillation. The GC-MS spektra could not show a matching peak, ^{1}H- and ^{13}C-NMR spectra would provide more reliable data on the structure, but was not meassured.

The nucleophilic electron pair of amino-group in the valinol reagent adds to the electrophilic C-atom of the polar carbonyl group of the diethylcarbonate, resulting in the elimination of the ethanolate as a good leaving group. The deprotonation of the hydroxy group by ethanolate initiates the cyclocarbonylation shown in scheme (7) to give (S)-4-isopropyl-2-oxazolidinone (**9**).

Chiral auxiliaries such as the classical Evans oxazolidin-2-ones have been widely used in the synthesis of natural products and pharmacologically active compounds.[8]

Scheme 7: reaction mechanism.

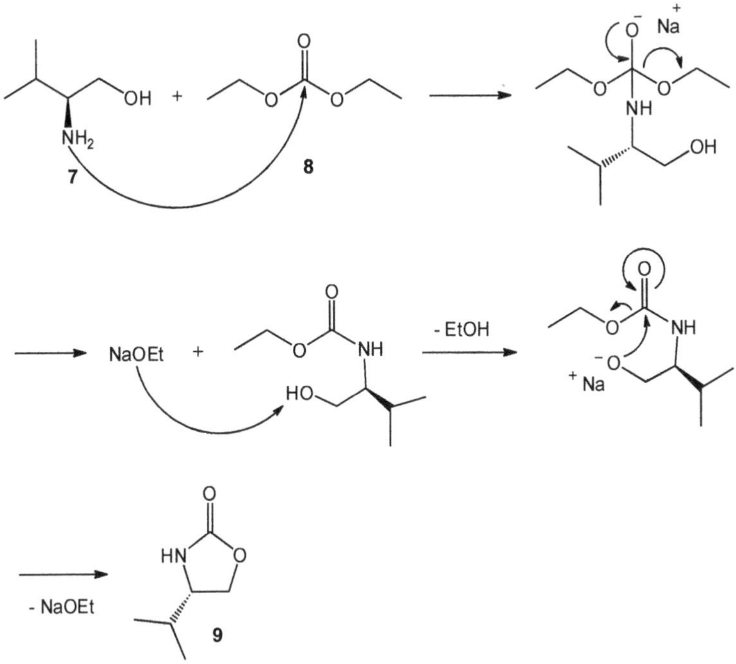

1.5 Synthesis of di-*tert*-butyl 1-(2,6-diisopropylphenyl)-hydrazine-1,2-dicarbo-xylate (11)[9]

A literature survey *via Reaxys* revealed 6 references in which this compound was described as the product. The first synthesis has been published by Klauber, Tetrahedron Lett., 1987: To turn the DIPP bromide into a Grignard needed for the addition of azodicarboxylate (**10**), 2,6-diisopropylphenylbromide was reacted with magnesium to give the *in situ* prepared 2,6-diisopropylphenylmagnesiumbromide (DIPPMgBr). This was reacted with azodicarboxylate (**10**) in THF at −78 °C to give di-*tert*-butyl-1-(2,6-diisopropylphenyl)-hydrazine-1,2-dicarboxylate (**11**) as dark orange viskose liquid in 98 % yield. Bredihhin, *Tetrahedron,* **2008**, 64, 6788–6793 is the best choice for its high yield.

Scheme 8: reaction equation.

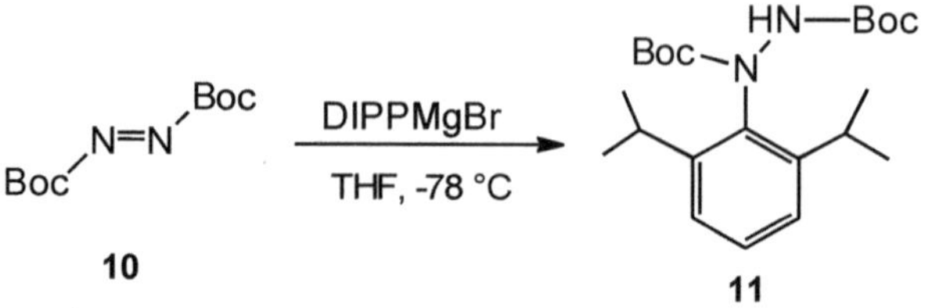

The reduction of azodicarboxylate (**10**) with the *in situ* prepared DIPPMgBr has given di-*tert*-butyl-1-(2,6-diisopropylphenyl)-hydrazine-1,2-dicarboxylate (**11**) as a dark orange viskose liquid in 98 % yield. This yield is very good and exceed that of litetartur[9] (84 %). GC-MS spectrum shows at t_R = 8.98 min., m/z 280.9 (3.0 %, [M − Boc]$^+$ calc.: 280.10), which can be assigned to the product. The messured melting point was 120 °C (lit.:[9] 121–122 °C), which indicates an acceptable purity, also the ^{1}H-, ^{13}C-NMR and the IR data (see experimental section and appendix) were consistent with the lit.[9].

For clarity, the mechanism has been formulated together with the synthesis of 1.6 (see scheme 10).

1.6 Synthesis of 2-(2,6-diisopropyl-phenyl)hydrazin-1-ium chloride (12)[9]

A literature survey *via Reaxys* revealed 6 references in which this compound was described as the product. The first synthesis has been published by Klauber, Tetrahedron Lett., 1987: Di-*tert*-butyl-1-(2,6-diisopropylphenyl)-hydrazine-1,2-dicarboxylate (**11**) was dissolved in a solution of HCl in dioxane and heated for 1 h at 60 °C to give the produkt 2-(2,6-diisopropylphenyl)hydrazin-1-ium chloride (**12**) as colourless needles in 51 % yield. Bredihhin,*Tetrahedron,* **2008**, 64, 6788–6793 is the best choice to the best of my knowledge due to the simplicity of the reaction.

Scheme 9: reaction equation.

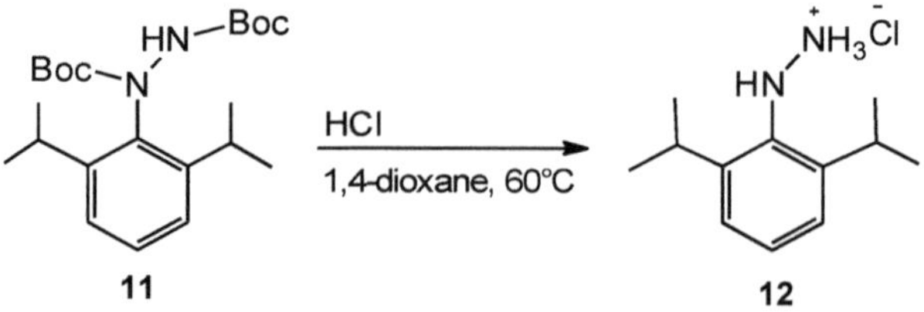

The elimination of both Boc-protecting groups from (**11**) has given 2-(2,6-diisopropyl-phenyl)hydrazin-1-ium chloride (**12**) as colourless needles in 51 % yield. This yield exceeds also that of litetartur[9] (32 %). The ^{1}H-NMR, ^{13}C-NMR spectra (see appendix) shows that the compound (**12**) was obtained pure. The GC-MS spectrum shows at t_R = 10.94 min. a peak with m/z 161.10 (90%, $[M - NH_4Cl]^+$ calc.: 161.30), which can be assigned as the Product fragmentation peak.

To form the Grignard a single electron transfer from magnesium to the bromide atom of 2,6-diisopropylphenylbromide gives the 2,6-diisopropylphenylradical, a bromide anion and a magnesium radical cation. The bromide anion binds to the magnesium radical to form magnesiummonobromide radical, now both radicals recombine to give the Grignard DIPPMgBr. The reaction with azodicarboxylate (**10**) and the following protonation with acetic acid gives the di-*tert*-butyl-1-(2,6-diisopropylphenyl)-hydrazine-1,2-dicarboxylate (**11**). To deprotect the hydrazin group and to obtain the product (**12**) HCl is added (see scheme 10).

Scheme 10: reaction mechanism.

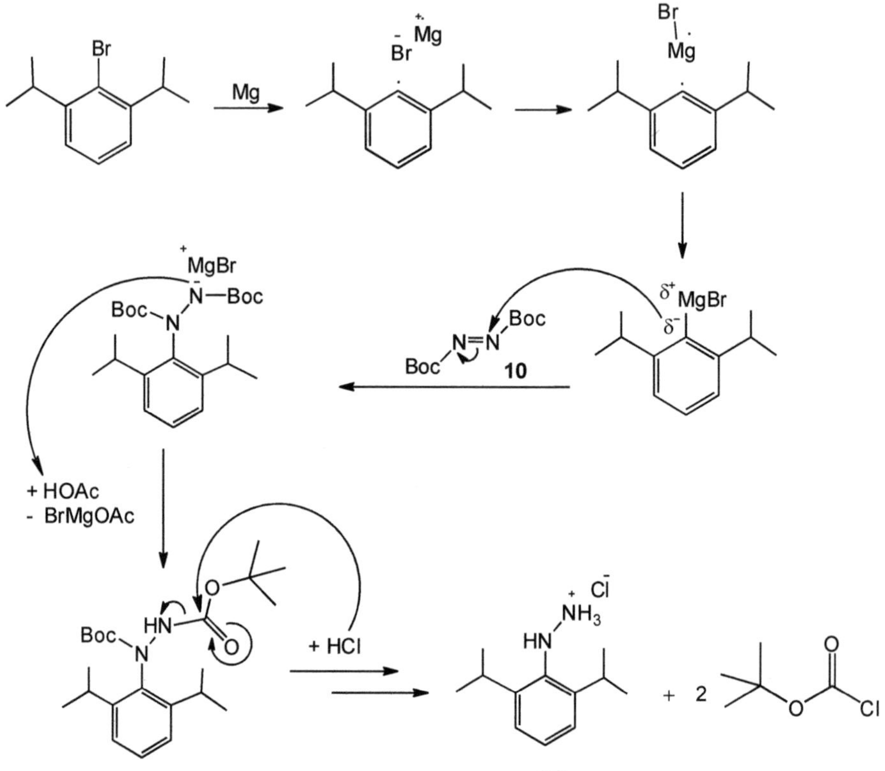

1.7 Synthesis of dimethyl *o*-tolylboronate (14)[10]

A literature survey *via Reaxys* revealed 9 references in which this compound was described as the product. The first synthesis has been published by Koning, Org. Biomol. Chem., 2004: To get the phenylboronic acid, 2-bromtoluone was reacted with BuLi and trimethylborate at −78 °C to give dimethyl-o-tolylboronate (**14**) as off-colourless crystalls in 98 % yield. Rakhi Pathak, Kantharuby Vandayar, Willem A. L. van Otterlo, Joseph P. Michael, Charles B. de Koning, *Org. Biomol. Chem.*, **2004**, 2, 3504 is the best choice due to its high yield, the simplicity of carrying it out and the short reaction time.

Scheme 11: reaction equation.

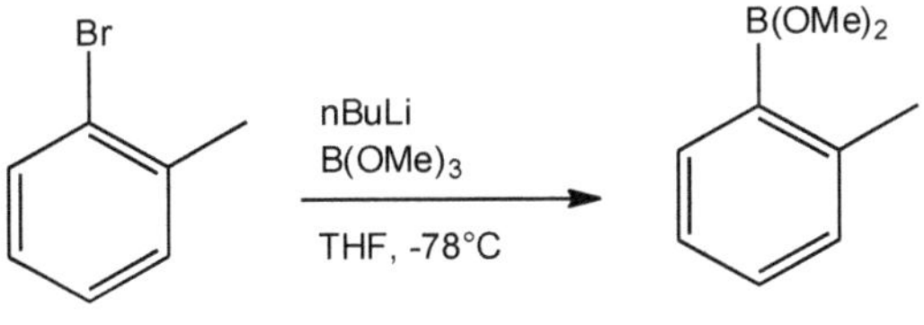

In this process, *n*-butyllithium was reacted with a phenylbromide (**13**) to provide the corresponding phenyllithium derivative, whereupon a borate was added under the formation of target product phenylboronic acid (**14**) in almost quantitative yield of 98 % as off-colourless crystalls, which exceed that of litetartur[9] (92 %). The GC-MS spactrum (see appendix) shows the peak at t_R = 8.93 min. m/z   91.10 (100%, $[M - B(OMe)_2]^+$ calc.: 91.13), which can be assigned to the product fragmentation peak.

The reaction between the bromobenzene (**13**) and BuLi provides the corresponding phenyllithium derivative. In the second step, trimethyl borate is introduced, which reacts with the phenyllithium intermediate that was produced during the first reaction step to provide the target product phenylboronic acid (**14**) after treating with HCl (see scheme 12).

Scheme 12: reaction mechnaism.

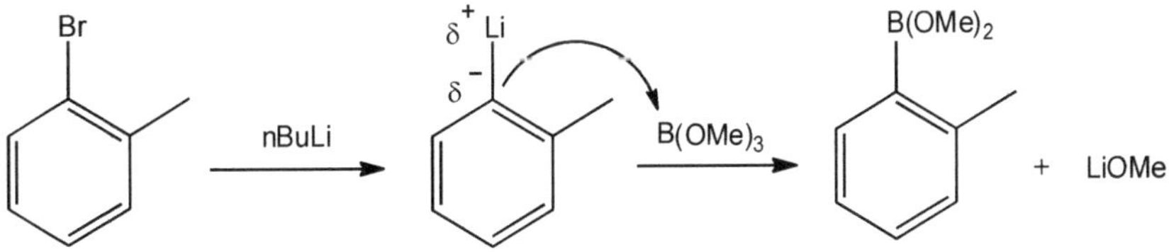

1.8 Synthesis of 1-methyl-4'-methoxybiphenyl (16)[11]

A literature survey *via Reaxys* revealed 6 references in which 1-Methyl-4'-methoxy-biphenyl (**16**) was described as the product. The first synthesis has been published by Goodson, Org. Synth., 2004: In a Suzuki coupling dimethyl-o-tolylboronate (**14**) was reacted with 1-iodo-4-methoxybenzene (**15**), using palldiumacetate as a catalyst and potassiumcarbonate as an additive in aceton/H_2O under Argon atmosphere and refluxed for 2 h at 70 °C to give 1-methyl-4'-methoxy-biphenyl (**16**) as a colourless solid in 49 % yield. Felix E. Goodson, *Org. Synth.*, **2004**, 10, 501 is the best choice for its short reaction time, mild conditions and greater catalytic turnover compared to other synthesis (0.02% catalyst is required for complete conversion).

Scheme 13: reaction equation.

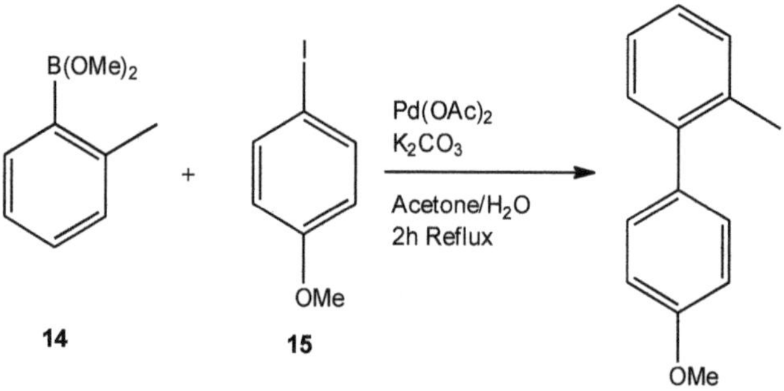

This a versatile method for synthesizing unsymmetrical biaryls has given by coupling of dimethyl-o-tolylboronate (**14**) and 1-iodo-4-methoxybenzene (**15**) 1-methyl-4'-methoxy-biphenyl (**16**) as a colourless solid in 49 % yield. The literature reports the product as a colourless oil and not as a colourless solid, so that it could not be destilled as recommended in the literature, besides the yield (49 %) is less than that reported in literature[11] (90 %). This low yield could be attributed to the fact that sometimes amounts of phenylated by-products are produced into the product mixture. This occurs during cross-coupling and represents a nonproductive consumption of aryl halides.[11] Another reason for the low yield could be also the deactivation of the catalyst.

The first step is the oxidative addition of palladium to the 1-iodo-4-methoxybenzene (**15**) to form the organo-palladium species (I–Pd–aryl). This step is the rate determining step in the catalytic cycle. The oxidative addition initially forms the *cis*-palladium complex, which rapidly isomerizes to the *trans*-complex. The displacement of the halide from I–Pd–aryl to give the more reactive organopalladium hydroxide (HO–Pd–aryl), because the Pd-O bond is more polar than a Pd-O-bond, thus the palladium hydroxide is more electrophilic (palladium is at the positive end of the dipole). Thereby facilitating the electrophilic transmetalation and forms with the boronate complex the organopalladium species (aryl–Pd–aryl). The next step is the *cis*/*trans*-isomerization for a better aryl-aryl-coupling. Reductive elimination of the desired product (**16**) restores the original palladium catalyst (see scheme 14)[12].

Scheme 14: reaction mechnism.

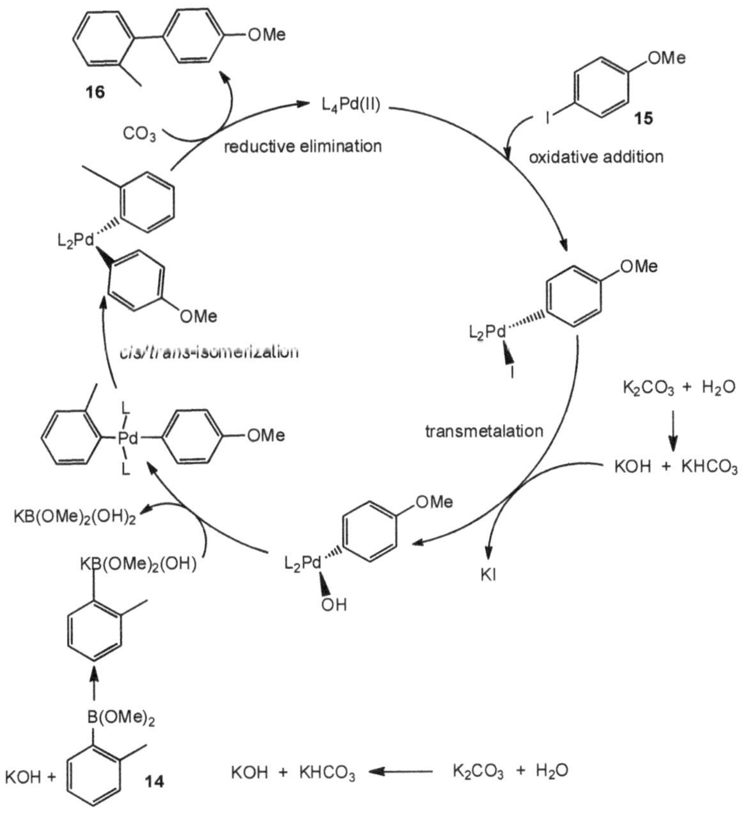

1.9 Synthesis of 2-iodobenzoic acid (18)[13]

A literature survey *via Reaxys* revealed 4 references in which this compound was described as the product. The first synthesis has been published by Tietze, Eicher., Reactions and Synthesis, 2007: In a Sandmeyer reaction 2-aminobenzoic acid (anthranilic acid) (**17**) reactes with sodium nitrite using potassium iodide as a catalyst in H_2O at $0-5°C$, followed by warming at 50 °C for 15 min. to give 2-iodobezoic acid (**18**) as yellow-orange needles in 90 % yield. Tietze, Eicher, Reactions and Synthesis, *Wiley-VCH*, **2007** is the best choice to the best of my knowledge due to its short reaction time and mild conditions.

Scheme 15: reaction equation.

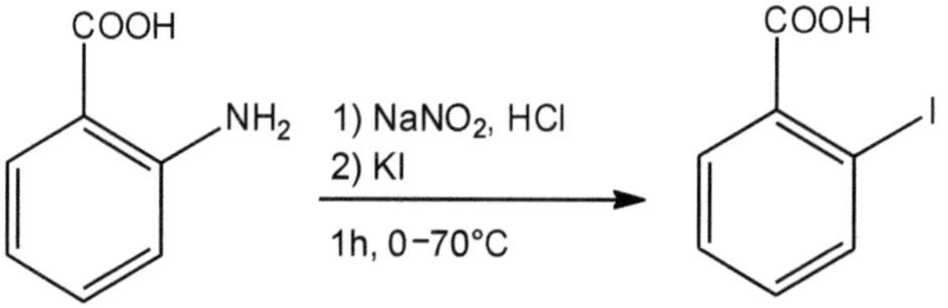

COOH

NH₂   1) NaNO₂, HCl   COOH

2) KI   I

1h, 0–70°C

17   **18**

Sandmeyer reaction is the name for the substitution of an amino group in aromatic compounds via their Diazonium salt by another nucleophile in this case iodide. 2-iodobenzoic acid (**18**) was obtained as yellow-orange needles in 90 % yield. This yield is good and exceed that reprorted in litetartur[13] (68 %). The retention factor R_f of the product 0.31 differs from that reported in lit. (0.39) and the melting point of 158 °C is consistent with literature (159 – 160 °C).

The nitrosation of primary aromatic amine of 2-aminobenzoic acid (**17**) with nitrous cation (NO^+) generated *in situ* from sodium nitrite ($NaNO_2$) and hydrochloric acid (HCl) leads to the diazonium salt after the nuceophilic attack by the amino group of the benzoic acid and the sequential formation of a water molekül. Via single-electron-transfer an electron ist transferred from iodide anion to the diazo group. Iodide radical adds to another iodide anion to give an iodide radical anion ($\cdot I_2^-$), which leads with its single electron to the elimination of a nitrogen molekül and the formation of a free aryl radical, that reacts with an iodide anion (I_2^-) to give 2-iodobenzoic acid (**18**)(see scheme 16).

Scheme 16: reaction mechanism.

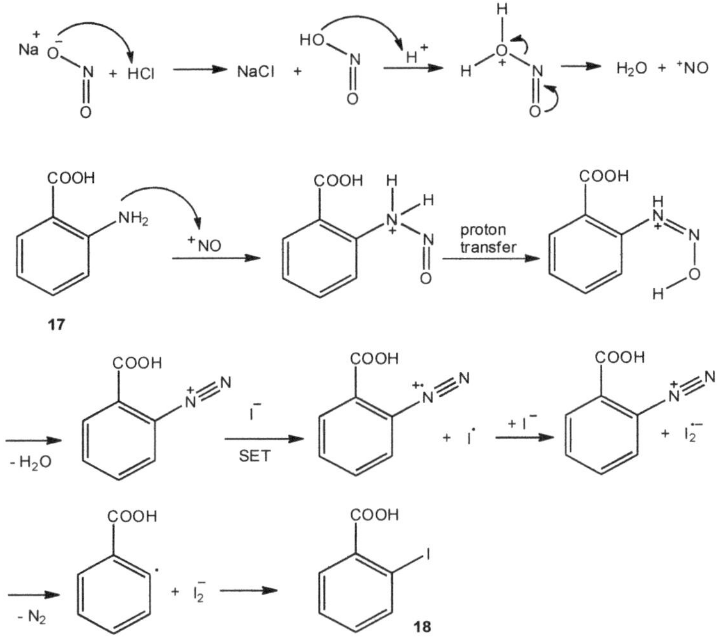

1.10 Synthesis of 1,1,2,2-tetraphenylethylen (20)[14]

A literature survey *via Reaxys* revealed 17 references in which this compound was described as the product. The first synthesis has been published by Mukaiyama, T, J. Chem. Lett., 1973: In a McMurry reaction benzophenone (**19**) reactes with titanium tetrachloride and zink in THF at 66 °C for 5 h to give 1,1,2,2-tetraphenyl ethylen (**20**) as a colourless solid in 60 % yield. Mukaiyama, T, J. Chem. Lett. **1973**, 1041-1044 is the best choice to the best of my knowledge due to its short reaction time and mild conditions.

Scheme 17: reaction equation.

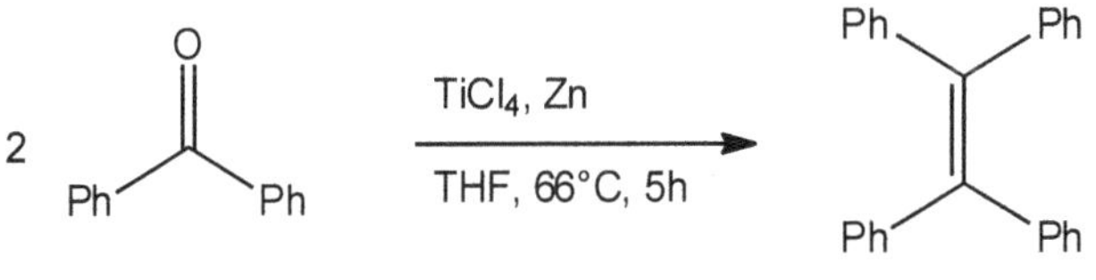

19 **20**

A single-electron transfer (SET) that is followed by a radical recombination has given 1,1,2,2-tetraphenyl ethylen (**20**) in 90 % yield. The yield is 7 % lower than that reported in the literarture (97 %). The GC-MS spectrum shows two peaks at t_R = 16.65 min. m/z 332.20 (100%, [M$^+$] calc.: 332,40) that can be assigned as the product. The second peak at t_R = 16.95 min. m/z 167.1 (100 %) [M-Na-2H] could be assiciated with diphenyl-methane (M = 168.24 g/mol), which might be a reduction side product. However further analytical data such as the messured melting point after the purification via column chromatography (Eluent cyclohexane) was 222 °C, which is consistent with the literature[15] (222 – 223 °C).

The McMurry reaction is running in the first step analogous to the pinacol reaction. The zink reduces titanium tetrachloride from the oxidation state (IV) to (II), this leads to the formation of titanium dichloride and zink dichloride. In a first step, a pinakolate is formed, in the second step the Pinakolate is deoxygenated to an alken. the driving force for this reaction is the formation of the strong O-Ti bond (see scheme 18).

Scheme 18: reaction mechanism.

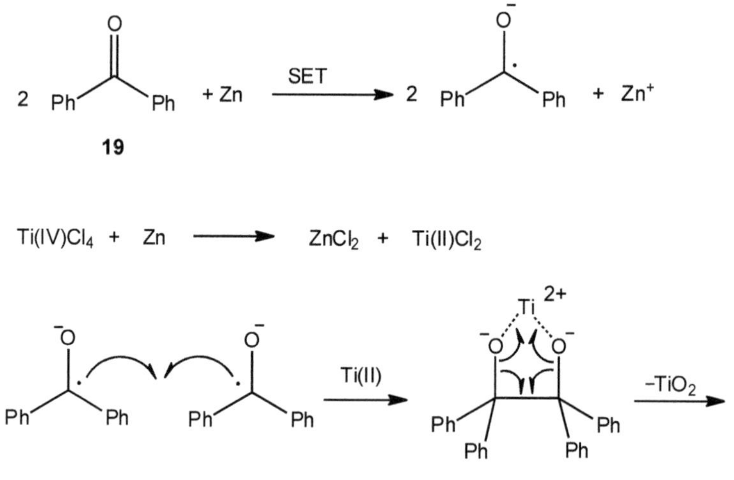

1.11 Synthesis of 5-hydroxy-3-oxo-5-phenylpentanoic acid methyl ester (23)[16]

A literature survey *via Reaxys* revealed 2 references in which this compound was described as the product. The first synthesis has been published by Paul. A. Clark and coworkers, Org. Biomol. Chem., 2005: Methyl acetoacetate (**21**) was reacted with sodium hydride and BuLi at room temperature, then benzaldehyde (**22**) was added at −78 °C. After 2 h of stirring at about −10 °C the *beta*-ketoester 5-hydroxy-3-oxo-5-phenyl-pentanoic acid methyl ester (**23**) as yellowish oil in 50 % yield was obtained. Paul A. Clarke and coworkers, **2005**, *Org. Biomol. Chem.*, 3551 - 3563 is the best choice to the best of my knowledge due to its high yield.

Scheme 19: reaction equation.

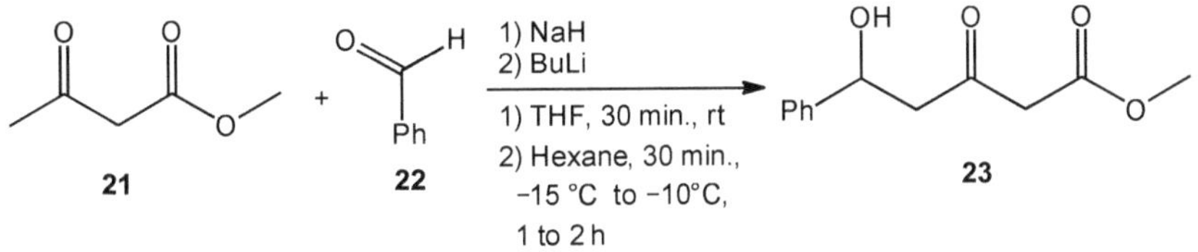

With the use of Weiler dianion chemistry of methyl acetoacetate (**21**) and the nucleophilic addition of benzaldehyde (**22**) the δ-hydroxy β-ketoesters 5-hydroxy-3-oxo-5-phenyl-pentanoic acid methyl ester (**23**) as aldol product in 50 % yield was obtained. The yield is below that reported in litetartur of 92 %. The cause therefor presumably could be the reaction between the deprotonated methyl acetoacetate (**21**) and the not deprotonated, although first hydride sodium was submitted and methyl acetoacetate (**21**) was added slowly dropwise so that during the deprotonation just a small concentration of (**21**) is present. However further analytical data such as the retention factor R_f 0.34 (Heptane : EtOAc, 2 : 1) was consistent with the literature[16] (0.33).

The difference in reactivity of the α- and γ-positions of the β-ketoester (**21**) has made it possible to deprotonate it in the α position with a base like NaH that is not so strong resulting in a double bond in the alpha position. Now to generate the so called Weiler dianion the stronger base BuLi was used to deprotonate the less reaktive γ-position resulting in a second double bond in the γ position, that undergoes a nucleophilic attack on the benzaldehyde (**22**). The final addition of water gives the δ-hydroxy β-ketoesters 5-hydroxy-3-oxo-5-phenyl-pentanoic acid methyl ester (**23**) (see scheme 20).

Scheme 20: reaction mechanism.

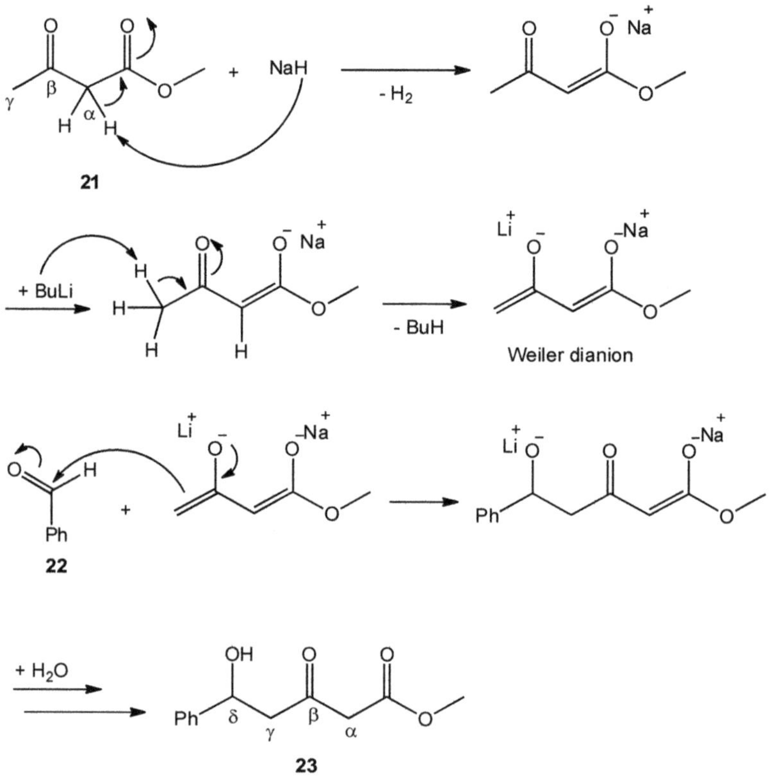

The δ-hydroxy β-ketoester 5-hydroxy-3-oxo-5-phenyl-pentanoic acid methyl ester (**23**) can be utilised in the Maitland - Japp reaction, instead of a ketone, as the building block for the tetrahydropyran (THP) ring. Taking advantage of the marked difference in reactivity of the α- and γ-positions of a δ-hydroxy β-ketoester would allow for the synthesis of unsymmetrical pyran products.[16]

2 Experimental Section

2.1 General procedure

2.1.1 Working method for air and moisture sensitive chemicals

For the processing of air- and moisture-sensitive substances, Schlenk technique has been used. As an inert atmosphere for reactions argon was used.

Chemicals were perchased at Sigma Aldrich, Merck AG, Acros Organics, Fulka Chemika and Alpha Aesar and were used without further purification.

The compounds prepared were stored in the refrigerator in the dark, and if necessary under a protective atmosphere.

2.1.2 Solvents

When working with air and moisture sensitive reagents freshly distilled solvents which were stored under inert gas were used. Drying agents for destillation are reported in table 1.

Table 1: The solvents used and the drying agents.

solvent	drying agent
toluene	Na
THF	Na
MeOH	Mg
DCM	CaH_2

2.1.3 Chromatographic Methods

TLC: Coated (silica gel 60 F254) aluminium plates, company *Merck* with 0.20 mm Kiesel gel coat fluorescent indicator. The eluent which was used is reported in the experimental section. Spot detection is carried out by fluorescence quenching at a wavelength of 254 nm.

$KMnO_4$ reagent:

In the case of compounds which do not have a conjugated π-system, the detection of the substance spots was performed on the TLC plates by immersion in the dyers reagent potassium permanganate and then heating with a hot air blower. Composition: $KMnO_4$

(3.00 g), KOH (0.30 g), K_2CO_3 (20.00 g) in 300.00 mL H_2O.

Column: Silica gel 60 (230 – 400 Mesh, company *Merck*)

2.1.4 Analytical methods

NMR: *Bruker* DPX 300 (300 MHz), chemical shifts δ in ppm, standard of spectrometer was used, abbreviations of fine structure of spectra: s = singlet, d = doublet, dd = double doublet, t = triplet, q = quartet, m = multiplet, br = broad. The chemical shift is followed by the specification of multiplicity, coupling constants, number of atoms and assignment of atoms written in parenthesis.

GC-MS: *Agilent* HP6890 system and mass detector, carrier gas hydrogen with a pressure of 1.20 bar and a flow velocity of 30.00 mL/min, capillary column: Optima 1 MS (30.00 x 0.25 mm) from company *Macherey-Nagel.*

m.p.: B-545 melting point apparatus, company *Büchi*, measurement in open capillaries.

IR: Nicolet 380 FT – IR, company *Thermo Fisher Scientific.*

2.1.5 Miscellaneous information

Information about gloveboxes, melting-point apparatus, work under inert atmosphere, kugelrohr distillation (a.k.a. "bulb-to-bulb distillation") etc.

2.2 Synthesis of *N*-(benzyloxycarbonyl)-(*S*)-proline (3)[1]

Scheme 21: reaction equation.

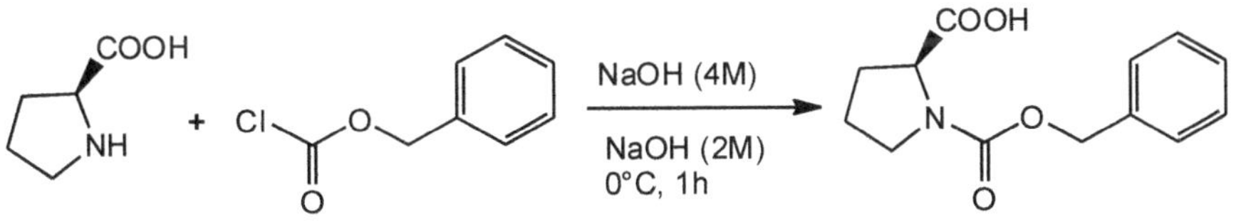

A 500 mL, threenecked, round-bottomed flash was equipped with two pressure equalizing dropping funnels (50.00 and 100.00 mL), a thermometer, and a magnetic stirrer. The flask was charged with a 2 M aqueous solution of sodium hydroxide (1.04 g/cm³, 100.00 mL, 0.20 mol) and cooled with an ice-salt bath. To the aqueous solution (*S*)-proline (23.00 g, 0.20 mol, Sigma Aldrich) was added with stirring. To the resulting solution benzyl chloroformate (1.20 g/cm³, 36.40 mL, 0.240 mol, Sigma Aldrich) and 4 M aqueous solution of sodium hydroxide (1.09 g/cm³, 70.00 mL, 0.28 mol) were added over 1 h with vigorous stirring and cooled to 0 to − 5 °C. The reaction mixture was stirred for another 1 h at 0 to − 5°C and then washed with ethyl ether (2 X 50.00 mL). The aqueous solution was acidified to pH 2.00 (ice bath cooling) by a dropwise addition of 6 M hydrochloric acid. The resulting mixture was saturated with sodium sulfate and extracted with ethyl acetate (3 X 100.00 mL). The extracts were combined, dried over anhydrous sodium sulfate, and evaporated under reduced pressure to give a colorless oil. The crystallization of the resulting mixture was induced by dissolving it in 20.00 mL of ethyl acetate, dilution with 100.00 mL of petroleum ether, cooling and scratching with a glass rod. The crystals were collected by filtration and washed with 20.00 mL of petroleum ether. After drying in vacuum colorless crystals of *N*-(benzyloxycarbonyl)-(*S*)-proline (**3**) were obtained.

Yield: 37.90 g (0.15 mol, 75 %, lit.:[1] 96 %)

mp: 72 °C (lit.:[1] 69 − 74 °C)

TLC: R_f = 0.70 (DCM/MeOH/Acetic acid 10 : 1 : 0.1,

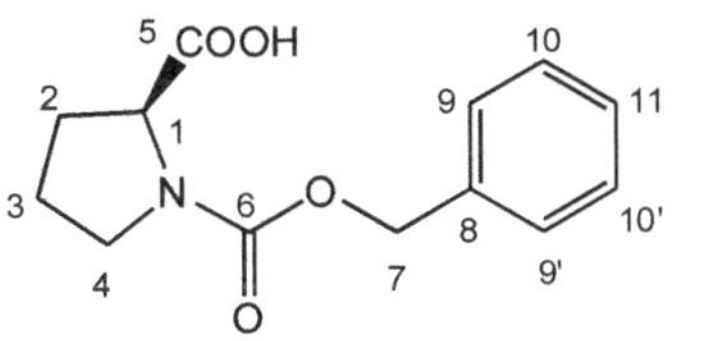

Scheme 22: numbering system of *N*-(benzyloxycarbonyl)-(*S*)-proline (**3**).

IR: $[cm^{-1}]$ = 2965.70 (s, v_{CH_2}), 1754.80 (ss, v_{CO}, carboxylic acid benzylester), 1645.60 (s, v_{CO}, primary, secondary carboxylic acid amide), 1187.50 (m, v_{C-OR}, ester), 1361.30 (w, v_{CN}, Amide).

2.3 Synthesis of N-(benzyloxycarbony1)-(S)-proline methyl ester[1]

Scheme 23: reaction equation.

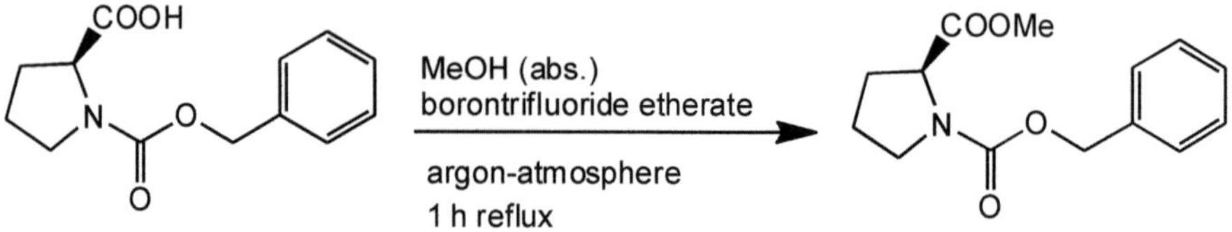

A 1 L, one-necked, round-bottomed flask equipped with a magnetic stirrer and a Liebig condenser fitted with a rubber septum was charged with N-(benzyl-oxycarbony1)-(S)-proline (33.70 g, 0.14 mol) and anhydrous methanol (0.79 g/cm³, 400 mL, 9.86 mol), and the contents were placed under dry nitrogen. After the addition of boron trifluoride etherate (1.15 g/cm³, 24.68 mL, 0.20 mol) with stirring, the solution was heated at reflux for 1 h. Solvent was removed under reduced pressure and the residue was vigorously stirred with 200.00 mL of ice-water and extracted with ethyl acetate (3 X 100.00 mL). The combined organic layers were successively washed with brine, 1 M aqueous sodium bicarbonate solution and brine and dried over anhydrous sodium sulfate. After the removal of the solvent under reduced pressure, the residual colorless oil was dried by twice dissolving in 100.00 mL of dry toluene and removing solvent under reduced pressure to give methyl ester as a colorless oil.

Yield: 4.00 g (0.015 mol, 95 %, lit.:[1] 85 %)

TLC: R_f = 0.43 (Cyclohexan/EtOAc, 2:1).

GC-MS: (t_R = 14.56 min., EtOAc) m/z (%): 263.1 (4%, [M]$^+$ calc.: 263.29), 77.0 (3 %) (phenylic cleavage), 91 (100 %) (benzylic cleavage).

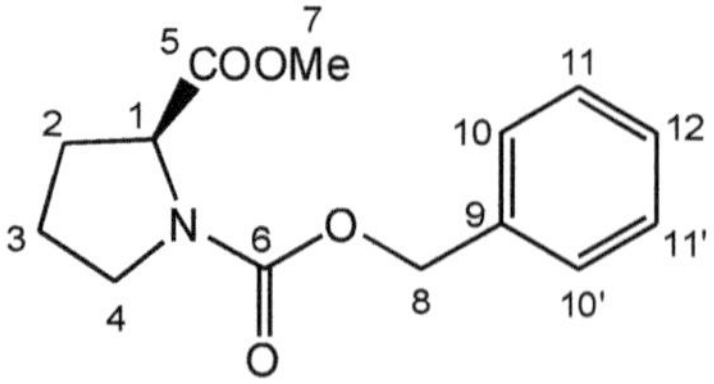

Scheme 24: numbering system of N-(benz-yloxy-carbony1)-(S)-proline methyl ester (**5**).

IR: [cm^{-1}] = 3033.1 (w, v_{CH}), 2952.9 (w, v_{CH_2}), 2880.2 (w, v_{OCH_3}), 1745.8 (m, v_{CO}, carboxylic acid benzylester), 1655.0 (s, v_{CO}, primary, secondary carboxylic acid amide), 1170.3 (w, v_{COR}, ester), 1353.6 (w, v_{CN}, Amide).

2.4 Synthesis of (*S*)-2-amino-3-methylbutan-1-ol (7)[3]

Scheme 25: reaction equation.

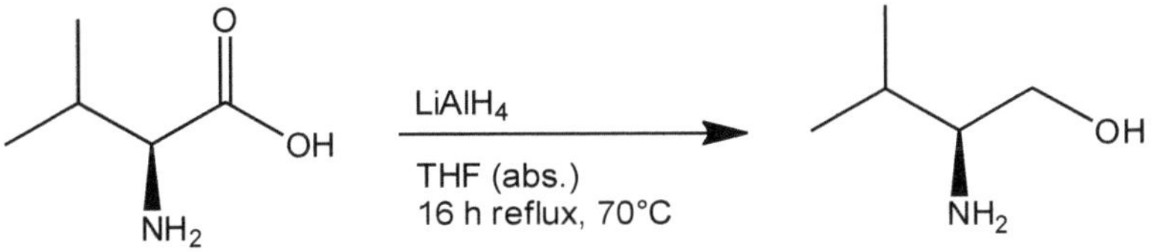

In a 100 mL three neck flask a solution of lithium aluminium hydride (0.46 g, 12.00 mmol, 1.40 equiv.) in dry THF (17.00 mL, 2.00 mL/mmol amino acid) was stirred and cooled to 0 °C in an ice bath. L-Valine (1.00 g, 8.53 mmol, 1.00 equiv.) was added and the mixture was stirred at 0°C for 1 h, warmed to room temperature, and stirred for an additional 1 h. The reaction mixture was then refluxed (70 °C) for 16 h. The solution was cooled in an ice bath and diluted slowly with diethyl ether (30 mL). Water (7.50 mL) and 15 % NaOH (2.50 mL) were added dropwise to the solution in succession. The mixture was filtered through Celite to remove aluminium salts. The Celite was washed with diethyl ether (2 × 15.00 mL). The filtrate was collected and the solvent was removed under reduced pressure affording the amino alcohol.

Yield: 0.90 g (7.70 mmol, 90 %, lit.:[3] 80 %)

TLC: R_f = 0.70 (DCM/MeOH/Et$_3$N, 10:1:0.1.

IR (neat) 3404.1 (w, v_{OH}), 3000 (w, v_{CH_3}), 2960 (w, v_{CH_2}), 1599 (w, v_{CO}, presumably rest of L-Valine),1337.5 (m, v_{CN}), 1044.8 (w, v_{CO}, primary alcohols) cm^{-1}.

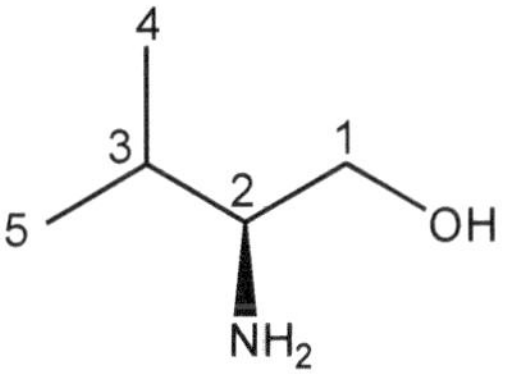

Scheme 26: numbering system of L-valinol (**7**).

Lit.[3] 3304 (s, v_{OH}), 2950 (w, v_{CH_2}), 2860 (s), 1590 (s), 1465 (m), 1385 (s), 1365 (m, v_{CN}), 1020 (s) cm^{-1}.

GC-MS: (t_R = 12.56 min., EtOAc) m/z (%): 105.20 (22%, [M+2H]$^+$ calc.: 105.15), 206.80 (13), 160.0 (100), 119.00 (6), 91.2 (24), 76.80 (15), 65.1 (18), 55.1 (12).

2.5 Synthesis of (*S*)-4-isopropyloxazolidin-2-one (9)[7]

Scheme 27: reaction equation.

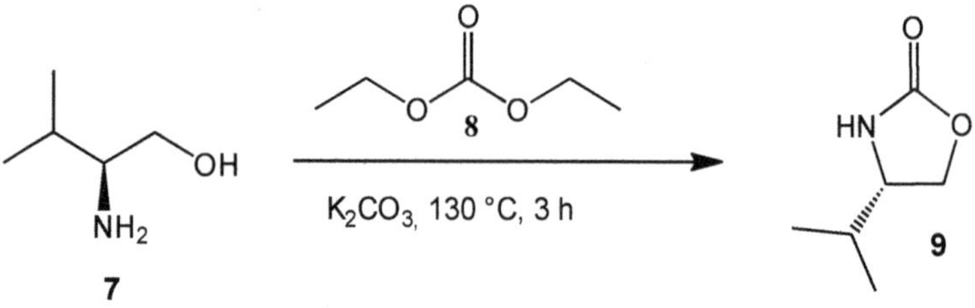

Potassium carbonate (0.28 g, 2.02 mmol) was added to a stirred solution of L-valinol (0.93 g/cm^3, 2.34 mL, 20.7 mmol) in diethylcarbonate (0.98 g/cm^3, 4.88 mL, 40.3 mmol) in a 50.00 mL round bottom flask. A short-path distillation apparatus was connected to this flask and the resulting solution was heated to 130 °C for 3 h while generated EtOH was removed by destilation. The residue was purified by aqueous extraction into dichloromethane (3 x 50.00 mL) using brine (50.00 mL) to give the oxazolidin-2-one (*S*)-2 (2.01 g, 80 %) as a colourless crystalline solid.

Yield: 0.80 g (6.20 mmol, 81 %, lit.:[7] 90 %)

TLC: R_f = 0.57 (DCM/MeOH), 10:1, **lit.:**[7] R_f = 0.3, Et$_2$O).

IR (CHCl$_3$) 3273.7 (m, v_{NH}), 1758.8 (s, v_{CO})
Lit. [7] 3263 (v_{NH}) and 1751 (v_{CO}) cm^{-1}.

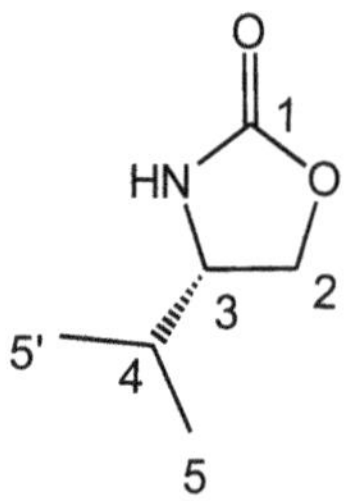

Scheme 28: numbering system of (*S*)-4-isopropyl-oxazolidin-2-one (9).

2.6 Synthesis of di-*tert*-butyl-1-(2,6-diisopropylphenyl)-hydrazine-1,2-dicar-boxylate (11)[9]

Scheme 29: reaction equation.

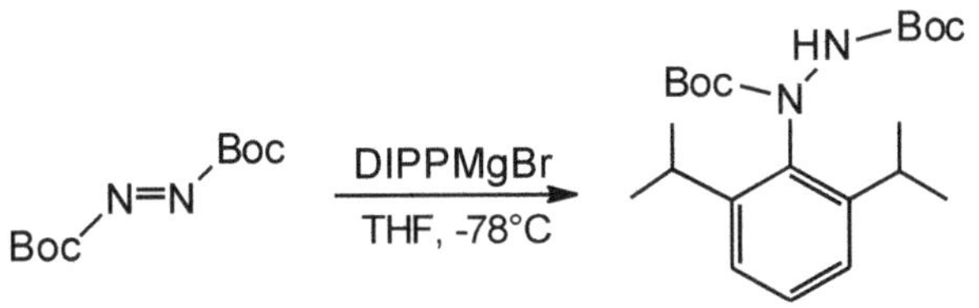

A small piece of I_2 was added to Mg (3.47 g, 143,00 mmol, 1.3 equiv.) in a heated 250 mL three neck flask under argon atmosphere. DIPP bromide (26.53 g, 110.00 mmol, 1.00 equiv.) and absolute THF (100.00 mL) were added. The mixture was heated to reflux for h before being allowed to cool to room rt. The solution of DIPPMgBr was added slowly to a solution of the azodicarboxylate (25.33 g, 110.00 mmol, 1.00 equiv.) in dry THF (300.00 mL) under argon atmosphere at −78 °C. After reaching the room temperature the reaction mixture was stirred over night. Acetic acid (1.05 g/cm³, 6.40 mL, 113 mmol, 1.03 equiv.) was added and THF was evaporated under reduced pressure. Water (250 mL) and Et_2O (250.00 mL) were edded to the residue. The aqueous layer was separated and extracted with Et_2O (2 X 250.00 mL). The combined organic layers dried over Na_2SO_4 and concentrated under reduced pressure to obtain the product.

Yield: 21.16 g (53.80 mmol, 98 %, lit.:[9] 84 %)

mp: 120 °C (lit.:[9] 121 – 122 °C)

IR: [cm⁻¹] = 2966.0 (s, v_{CH_3}),1704.1 (s, v_{CO}), 1154.3 (s, v_{COR}, ester), 1366.5 (w, v_{CN}, Amide).

¹H-NMR (300 MHz, CDCl₃): δ [ppm] = 9.81 (s, 2-3H, NH_{2-3}), 7.3 (t, $^3J_{HH}$ = 7.6 Hz, 1H, H-4), 7.10 (d, $^3J_{HH}$ = 7.6 Hz, 2H, H-3), 6.45 (s,1-2H, NH_{1-2}), 3.33 (sept. $^3J_{HH}$ = 6.6 Hz, 2H, H-5), 1.80 (m, 18H, H-9,12), 1.26 (m, 12H, H-6).

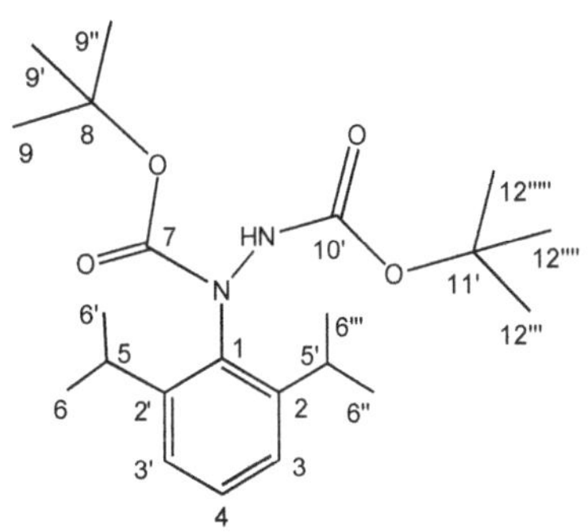

Scheme 30: numbering system of di-*tert*-butyl-1-(2,6-diisopropylphenyl)hydrazine-1,2-dicar-boxylate (**11**).

Lit. [7] (300 MHz, CDCl$_3$): δ [ppm] = 9.81 (s, 2H, NH$_2$), 7.28 (t, $^3J_{HH}$ = 7.5 Hz, 1H, H-4), 7.18 (d, $^3J_{HH}$ = 7.5 Hz, 2H, H-3), 6.61 (s, 1H, NH$_1$), 3.46 (sept. $^3J_{HH}$ = 6.6 Hz, 12H, H-6).

^{13}**C-NMR**: (300 MHz, CDCl$_3$): δ [ppm] = 176.81 (s, C7), 169.72 (s, C8), 152.44 (s, C2), 149.95 (s, C1), 128.92 (d, C4), 124.05 (d, C3), 28.13 (d, C5), 24.25 (q, C6, C9).

Lit. [7]: (300 MHz, CDCl$_3$): δ [ppm] = 146.60 (s, C2), 138.20 (s, C1), 128.60 (d, C4), 124.10 (d, C3), 27.60 (d, C5), 24.60 (q, C6).

GC-MS: (t_R = 8.98 min., EtOAc) m/z (%): 280.90 (3,0 %, [M − Boc]$^+$ calc.: 280.10), 207.00 (10), 162.10 (5), 147.20 (12), 134.90 (4), 120.10 (100), 106.10 (95), 91.00 (40), 77.00 (30), 63.00 (10), 50.70 (12).

2.7 Synthesis of 2-(2,6-diisopropyl-phenyl)hydrazin-1-ium chloride (12)[9]

Scheme 31: reaction equation.

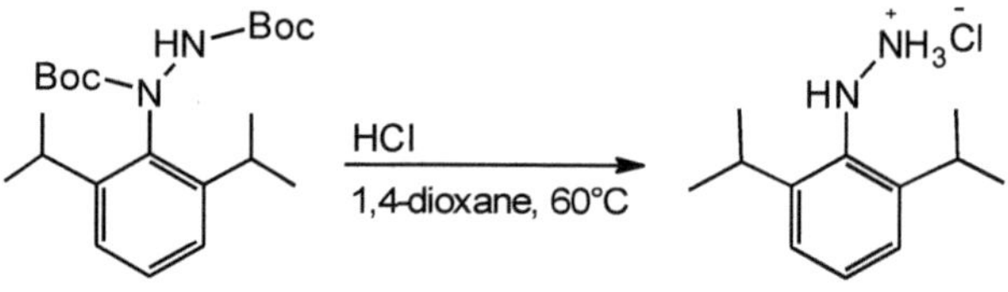

HN—Boc
Boc—N

HCl
1,4-dioxane, 60°C

NH$_3$Cl
HN

11 → 12

Di-*tert*-butyl-1-(2,6-diisopropylphenyl)-hydrazine-1,2-dicar-boxylate (23.00 g, 0.059 mol) (**11**) was dissolved in isopropanol (50.00 mL) and HCl solution (70.00 mL, in 1,4-dioxane) and heated to 60 °C for 1 h. The mixture cooled to 0°C and diluted with Et$_2$O (100.00 mL). A small amount of a colourless precipitate was filtered of. The filtrate was again warmed to 50 °C. After 10 min. a bulk of colourless precipitate was filtered of. The filtrate was diluted againe with Et$_2$O (100.00 mL) after some time againe an amount of the same precipitate was observed and filtered of. This process was repeated two more times untill no colourless precipitate was observed.

Yield: 6.90 g (0.03 mol, 51 %, lit.:[9] 32 %)

mp: 193 °C (lit.:[9] 192 – 194 °C)

^{1}**H-NMR** (300 MHz, CDCl$_3$): δ [ppm] = 9.70
(s, 2H, NH$_2$), 7.26 (t, $^{3}J_{HH}$ = 7.4 Hz, 1H,
H-4), 7.13 (d, $^{3}J_{HH}$ = 7.4 Hz, 2H, H-3), 5.89
(s, 1H, NH), 3.38 (sept. $^{3}J_{HH}$ = 6.7 Hz,
2H, H-5), 1.20 (d, $^{3}J_{HH}$ = 6.6 Hz, 12H, H-6).

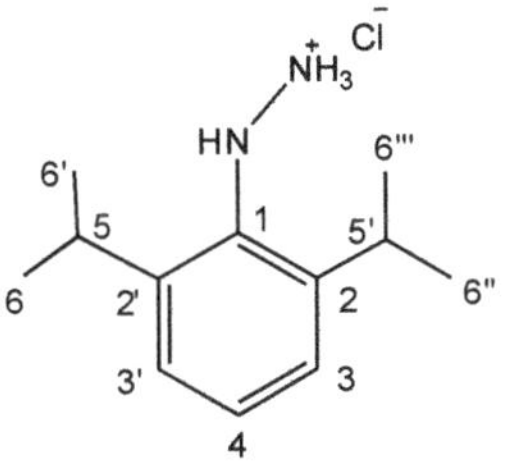

Scheme 32: numbering system
of 2-(2,6-diisopropylphenyl)
hydrazin-1-ium chloride (**12**).

Lit. [7] (300 MHz, CDCl₃): δ [ppm] = 9.81 (s, 2H, NH$_2$), 7.28 (t, $^{3}J_{HH}$ = 7.5 Hz, 1H, H-4), 7.18
(d, $^{3}J_{HH}$ = 7.5 Hz, 2H, H-3), 6.61 (s, 1H, NH$_1$), 3.46 (sept. $^{3}J_{HH}$ = 6.6 Hz, 2H, H-5), 1.20 (d,
$^{3}J_{HH}$ = 6.6 Hz, 12H, H-6).

^{13}**C-NMR**: (300 MHz, CDCl₃): δ [ppm] = 129.06 (d, C4), 124.21 (d, C3), 28.06 (d, C5),
24.25 (q, C6).

Lit. [7]: (300 MHz, CDCl₃): δ [ppm] = 128.6 (d, C4), 124.10 (d, C3), 27.60 (d, C5), 24.60
(q, C6).

GC-MS: (t_R = 10.94 min., EtOAc) m/z (%): 161.10 (90%, [M – NH$_4$Cl]$^+$ calc.: 161.30),
206.70 (93), 207.00 (10), 133.00 (100), 120.0 (50), 91.10 (70), 78.30 (67), 61.00 (55),
52.10 (65).

2.8 Synthesis of dimethyl-*o*-tolylboronate (14)[10]

Scheme 33: reaction equation.

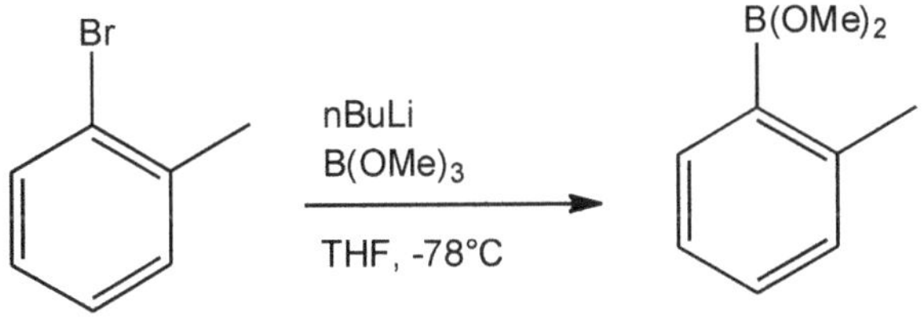

*n*BuLi (0.66 g/cm³, 4.20 mL, 6.60 mmol, 1.6 M in Hexane) was added dropwise to a solution of 2-bromotoluene (1.10 g, 6.40 mmol) in THF (30.00 mL) at −78 °C, resulting in a colourless suspension. The reaction mixture was stirred for 30 min. at −78 °C, then $B(OMe)_3$ (0.93 g/cm³, 1.83 mL, 16.40 mmol) was added. The resulting mixture was stirred at −78 °C for a further 30 min. and then allowed to warm to rt. The reaction mixture was acidified with 10 % aq. HCl solution and extracted with Et_2O (3 X 60.00 mL). The organic layer was then dried with $MgSO_4$ and concentrated under vacuum to afford an off-colourless crystalline material, dimethyl *o*-tolylboronate (**14**).

Yield: 0.85 g (5.30 mmol, 98 %, lit.:[10] 92 %)

mp: 71 °C (lit.:[10] 72 °C)

TLC: R_f = 0.40 (cyclohexane/EtOAc, 1:2,

GC-MS: (t_R = 8.93 min., EtOAc) m/z (%) 91.10 (100 %, $[M - B(OMe)_2]^+$ calc.: 91.13), 207.00 (10), 170.00 (30), 77.80 (12) phenylic cleavage, 64.80 (30), 51.10 (8).

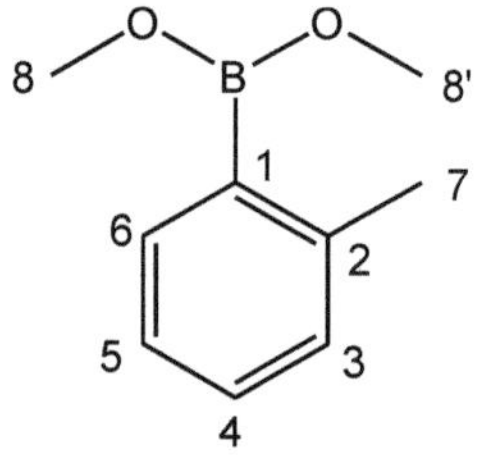

Scheme 34: Numbering system of dimethyl-*o*-tolylboronate (**14**).

2.9 Synthesis of 4-methoxy-2'-methylbiphenyl (16)[11]

Scheme 35: reaction equation.

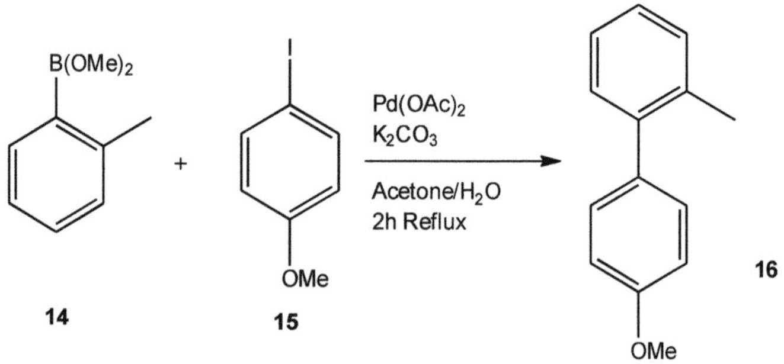

Dimethyl *o*-tolylboronate (**14**) (0.70 g, 4.91 mmol), 4-iodoanisole (1.10 g, 5.00 mmol), and acetone (15.00 mL) are combined in a 100 mL, three-necked flask equipped with an efficient stirbar, two stoppers, and a reflux condenser attached to a gas-flow adapter with a stopcock. Potassium carbonate (1.70 g, 12.00 mmol), is dissolved in water (15.00 mL) in a separate 25 mL Schlenk flask. In a third 25 mL Schlenk flask palladium acetate (0.22 mg, 13.30 μmol, 0.013 %) is dissolved in acetone (1.00 mL). All three flasks are then horoughly degassed by 15 min. argon inlet. Under an argon flow, one of the stoppers on the three-necked flask is replaced with a rubber septum, and the carbonate and catalyst solutions are added via cannula to form a biphasic mixture. The top layer turns brown upon addition of the catalyst. The septum is replaced with the glass stopper and an additional 10 min. argon inlet is applied. After 2 h at reflux the heat source is removed and the reaction is allowed to cool down. By this time the brown color has faded and the reaction is a triphasic mixture with copious amounts of palladium black (a black precipitate of elemental palladium, which forms on decomposition of palladium complexes) floating between the layers. The reaction is transferred to a 0.50 L separatory funnel and extracted into diethyl ether (3 × 10.00 mL). The organic layers are combined, washed with water (1 × 10.00 mL) saturated with sodium chloride, and dried over magnesium sulfate. Solvent is removed with a rotary evaporator and a colourless solid was obtained.

Yield: 0.48 g (2.42 mmol, 49 %, lit.:[11] 90 %)

TLC: R_f = 0.35 (Cyclohexane/EtOAc 2:1).

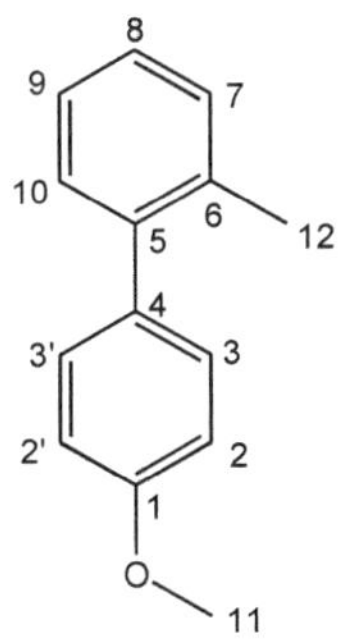

Scheme 36: numbering system of 4-methoxy-2'-methylbiphenyl (**16**).

31

2.10 Synthesis of 2-iodo benzoic acid (18)[13]

Scheme 37: reaction equation.

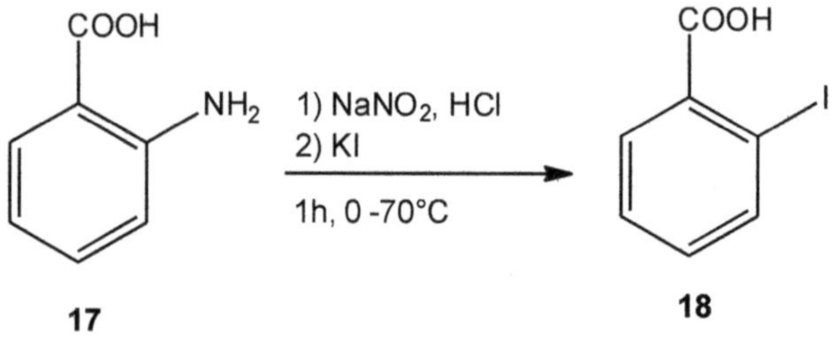

Under an argon atmosphere, anthranilic acid (3.42 g, 24.90 mmol) is dissolved in a mixture of H_2O (25.00 mL) and concentrated HCl (6.25 mL). The solution was cooled to $0-5°C$, and a solution of sodium nitrite (1.77 g, 25.70 mmol) in H_2O (5.00 mL) was added dropweise, while the temperature did not exceed 5 °C. The resulting mixture was stirred for 5 min. and then a solution of potassium iodide (4.27 g, 25.70 mmol) in H_2O (6.50 mL) was added. Stirring was continued for 5 min. without cooling, then the solution was warmed to 50 °C causing rapid gas evolution and the formation of a brown precipitate.

After 15 min. at 50 °C, the temperature was increased to 70 °C for 10 min., and then the solution was cooled with an ice bath. Sodium thiosulfate was added to desactivate iodine excess, and the precipitate was collected by filtration and washed with ice water (3 X 20.00 mL). The solid was dissolved in hot EtOH (17.50 mL) and treated three times with charcoal. The charcoal was filtered off, and the final filtrate was diluted with hot H_2O (8.00 mL) and heated to reflux. Cold H_2O (10.00 mL) was added, and the solution was left in a refrigerator to give yellow-orange needles.

Yield: 5.60 g (22.50 mmol, 90 %, lit.:[13] 68 %)

TLC: R_f = 0.313 (n Pentane/EtOAc/acetic acid 4:1:1, lit.:[13] R_f = 0.39).

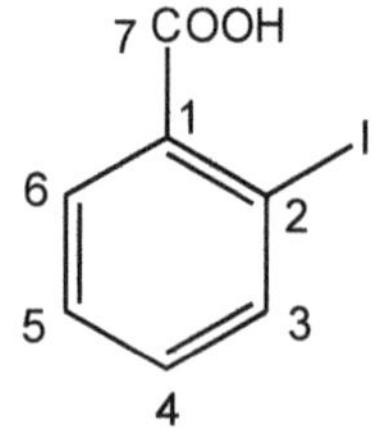

Scheme 38: numbering system of 2-iodobenzoic acid (18).

2.11 Synthesis of 1,1,2,2-tetraphenylethylene (20)[14]

Scheme 39: reaction equation.

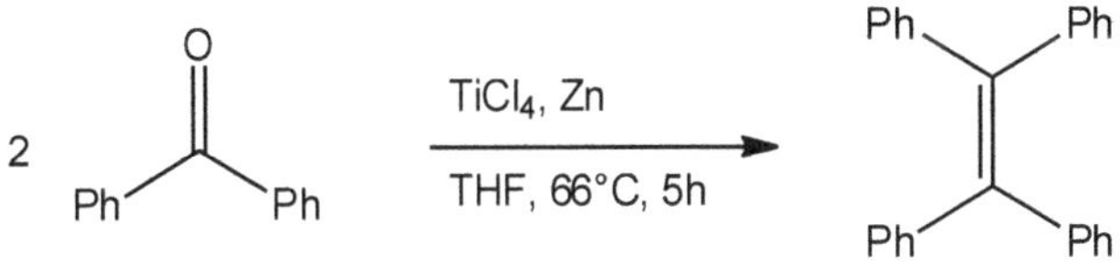

A suspension of zinc powder (0.59 g, 9.00 mmol) in abs. THF (10.00 mL) was added slowly into amixture of benzophenone (0.55 g, 3.00 mmol) and $TiCl_4$ (1.73 g/cm^3, 0.50 mL, 4.5 mmol) in THF (20.00 mL) at –10°C under argon. The yellow solution immediatley changed to purple and then turned dark brown. The reaction mixture was stirred for 5 h at 70 °C, followed by alkaline hydrolysis with K_2CO_3 solution (10 %), ether extraction (3 X 20.00 mL) and drying of the combined organic layers over $MgSO_4$. The Solvent is removed with a rotary evaporator. 1,1,2,2-Tetraphenyl-ethylene (**20**) was obtained as a colourless solid.

Yield: 0.60 g (1.80 mmol, 60 %, lit.:[14] 97 %)

mp: 222 °C (lit.:[14] 222 – 224 °C)

GC-MS: (t_R = 16.65 min., EtOAc) m/z (%): 332.2 (100 %, [M]$^+$ calc.: 332.40), 317,20 (5), 302.00 (5), 289.00 (5), 276.10 (3), 253.10 (45), 239.10 (17), 226.00 (5), 207.00 (4), 191.20 (2), 178.00 (9), 165.10 (15), 151.00 (7), 126.10 (7), 113.00 (4), 91.10 (3), 77.10 (6), 51.00 (6).

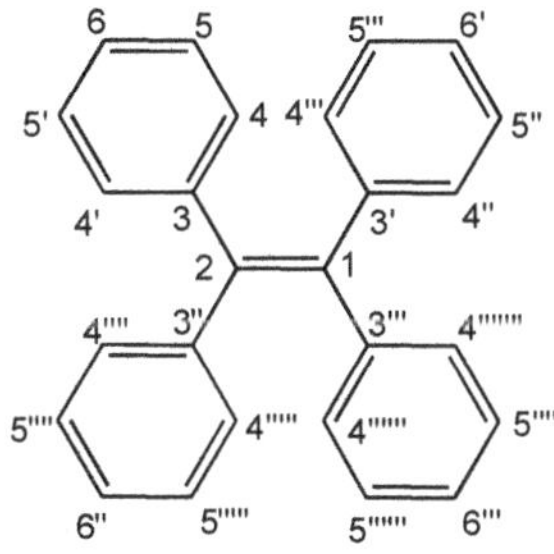

Scheme 40: numbering system of 1,1,2,2-tetraphenylethylene (**20**).

2.12 Synthesis of 5-hydroxy-3-oxo-5-phenyl-pentanoic acid methyl ester (23) [16]

Scheme 41: reaction equation.

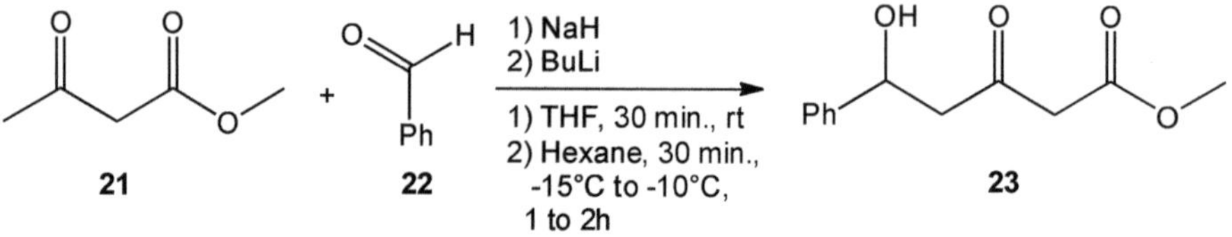

To sodium hydride (0.41 g, 10.34 mmol) in THF (20.00 mL) at 0 °C was added methyl acetoacetate (1.02 g/cm^3, 1.14 mL, 10.00 mmol) slowly over 5 min. during which time gas evolution was observed. The colourless solution was stirred for 10 min. at 0 °C and then BuLi (0.66 g/cm^3, 6.25 mL, 10.00 mmol, 1.6 M in hexanes) was added. The light yellow solution was stirred at room temperature for 20 min. and was then cooled in an acetone/dry ice bath. Once the internal temperature had reached −78 °C the aldehyde (1.05 g/cm^3, 1.10 mL, 10.00 mmol) was added over a 5 min. period. The solution was kept at −78 °C for 5 min. and was then warmed to room temperature over 30 min. The light yellow solution was stirred for 30 min. and then H_2O (10.00 ml) was added. The mixture was extracted with EtOAc (50.00 mL) and washed with 5 % $NaHCO_3$ (3 X 30 mL) and brine (2 X 30 mL), dried over $MgSO_4$, and concentrated in vacuo to give the aldol products (**23**) as yellow oil.

Yield: 1.10 g (5.00 mmol, 50 %, lit.:[16] 92 %)

TLC: R_f = 0.34 (n Heptane/EtOAc 2 : 1, lit.:[16] R_f = 0.33).

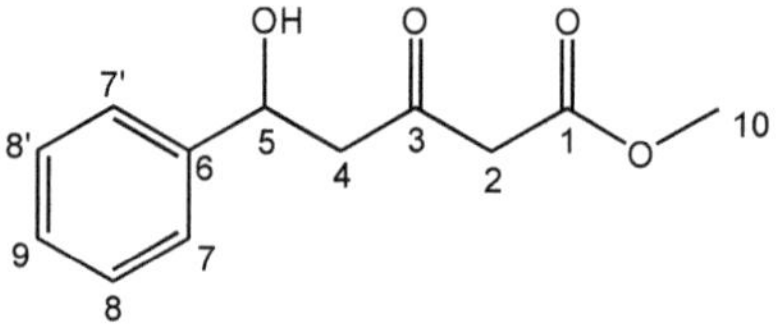

Scheme 42: numbering system of 5-hydroxy-3-oxo-5-phenyl-pentanoic acid methyl ester (**23**).

3 Appendix

3.1 List of abbreviations

equiv.	equivalent
BuLi	butyl lithium
°C	Ceslius
d	dublett
dd	double doublet
DCM	dichloromethane
dt	double triplett
δ	chemical shift
Et_2O	diethyl ether
EtOAc	ethyl acetate
$CDCl_3$	deutero chloroform
$CHCl_3$	chloroform
g	gramm
h	hour
HCl	hydrochloric acid
Hz	Hertz
IR	infra red
J	coupling constant
KOH	potassium hydroxide
L	liter
M	molarity
m	multiplett
MeOH	methanol
MHz	mega Hertz
mg	milligramm
$MgSO_4$	magnesiumsulfate
mL	milliliter
mmol	millimol
mp	melting point
MS	mass spectrum

Na	sodium
nm	nano meters
NMR	nuclear magnetic resonance
pH	potentia hydrogenii
ppm	parts per million
R_f	retention factor
rt	room temperature
q	quartett
s	singlett
t	triplett
THF	tetrahydrofurane

3.2 References

[1] E. J. Corey, Saizo Shibata, Raman K. Bakshi, „An efficient and catalytically enantio-selective route to (S)-(-)-phenyloxirane", *J. Org. Chem.*, Vol. 53, No. 12, **1988**, 2861 - 2863.

[3] M. Klein, K. Krainz, I. Namro, P. Diner and M. Grotli, Synthesis of Chiral 1,4-Disubstituted-1,2,3-Triazole derivatives from Amino Acids, *Molecules*, **2009**, 14, 5124.

[4] Welsh, J. T., Seper, K. W. Synthesis, *J. Org. Chem.*, **1988**, 53, 2991 - 2999.

[5] https://www2.chemistry.msu.edu/faculty/reusch/VirtTxtJml/crbacid2.htm.

[6] McManus, Helen A.; Guiry, Patrick J. "Recent Developments in the Application of Oxazoline-Containing Ligands in Asymmetric Catalysis". *Chemical Reviews* **104** (9): 4151 - 4202.

[7] David Benoit,a Elliot Coulbeck,b Jason Eamesb, and Majid Motevalli, „On the structure and chiroptical properties of (S)-4-isopropyl-oxazolidin-2-one", *Tetrahedron: Asymmetry* 19 (**2008**) 1068 - 1077.

[8] Farina, V.; Reeves, J.T.; Senanayake, C.H.; Song, J.J. Asymmetric synthesis of active pharmaceutical ingredients. *Chem. Rev.* **2006**, *106*, 2734 - 2793.

[9] Bredihhin, A., Mäeorg, U. Effective strategy for the systematic synthesis of hydrazine derivatives. *Tetrahedron,* **2008**, 64, 6788 - 6793.

[10] Rakhi Pathak, Kantharuby Vandayar, Willem A. L. van Otterlo, Joseph P. Michael, Manuel A. Fernandes and Charles B. de Koning, The synthesis of angularly fused polyaromatic compounds by using a light-assisted, base-mediated cyclization reaction, *Org. Biomol. Chem.*, **2004**, 2, 3504.

[11] Felix E. Goodson, Thomas I. Wallow, and Bruce M. Novak, accelerated Suzuki coupling via Ligandless Palladium Catalyst, *Org. Synth.*, **2004**, 10, 501.

[12] A. Suzuki, N. Miyaura, *Chemical Reviews*, **1995**, 95, 2457 - 2483.

[13] Tietze, Eicher, Reactions and Synthesis, *Wiley-VCH*, **2007**.

[14] Mukaiyama, T.; Sato, T.; Hanna, J. Chem. Lett. **1973**, 1041 - 1044.

[15] Irwin C. Lewis, T. Edstrom, Thermal Reactivity of Polynuclear Aromatic Hydrocarbons, *J. Org. Chem.*, **1963**, 28 (8), 2050 - 2057·

[16] Paul A. Clarke, The one-pot, multi-component contrauction of highly substituted tetrahydropyranones using the maitland-Japp raection, School of Chemistry, University of Nottingham, **2005**, *Org. Biomol. Chem.*, 3551 - 3563.

3.3 Spectra

3.3.1 *N*-(benzyloxycarbonyl)-(*S*)-proline (**3**)[1]

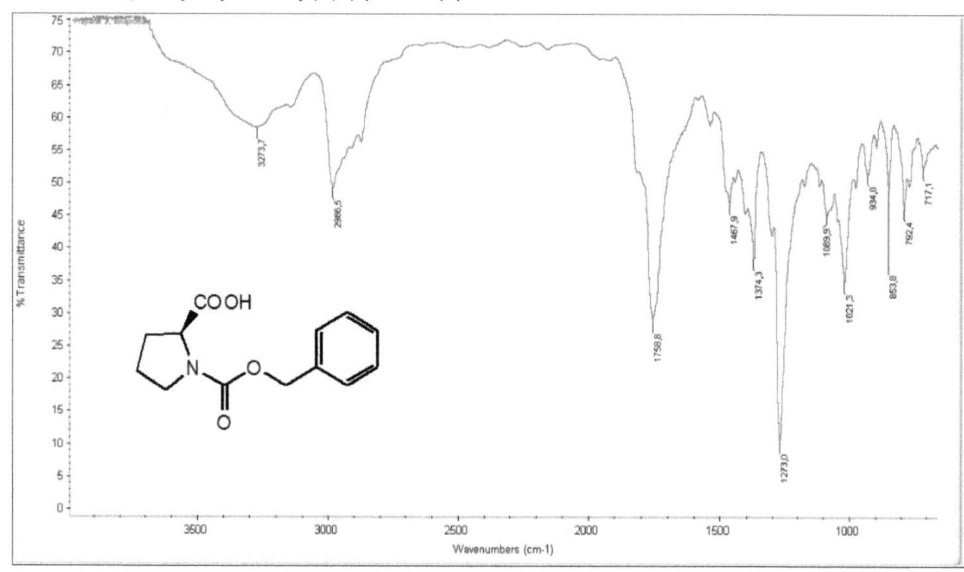

3.3.2 *N*-(benzyloxycarbonyl)-(*S*)-proline Methyl Ester (**5**)[1]

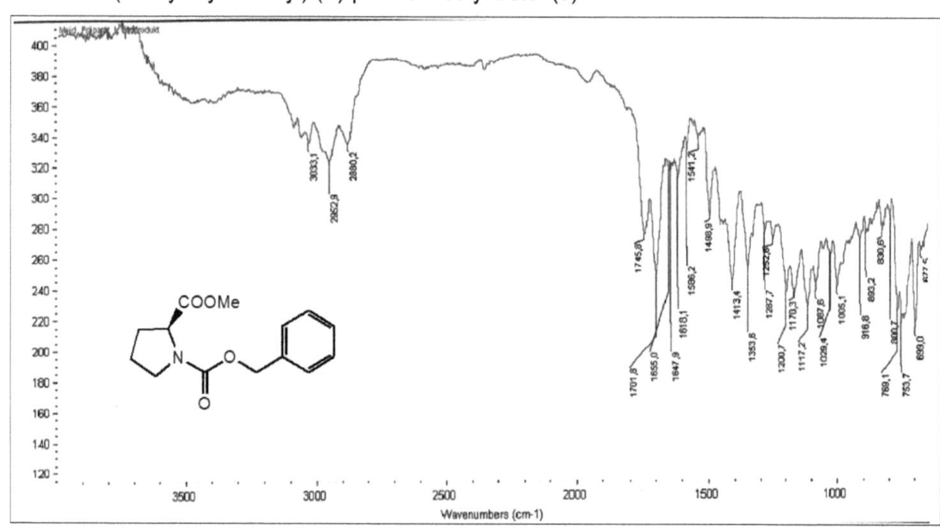

File   :C:\msdchem\1\DATA\DANOCSA2.D
Operator : CMS
Acquired : 9 Oct 2015   6:50   using AcqMethod STAND35.M
Instrument :   MSD 5975
Sample Name: DANOCSA2
Misc Info :
Vial Number: 26

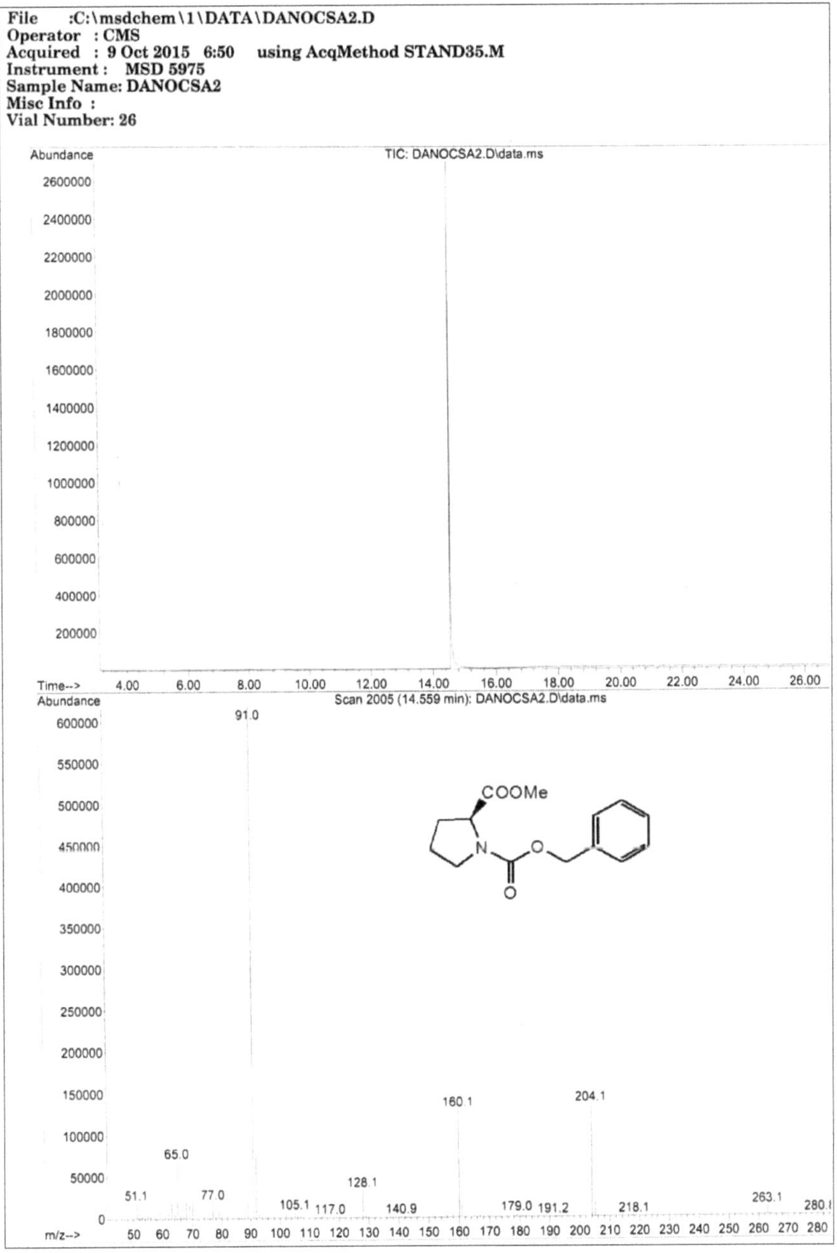

3.3.3 2-Amino-3-methylbutanol (Valinol) (**7**)[3]

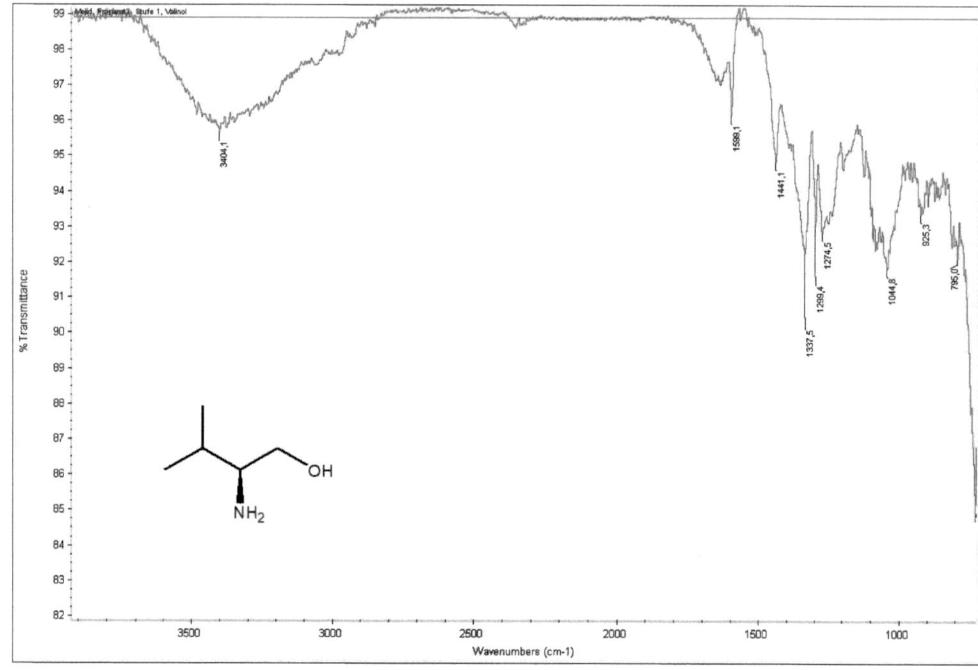

File   :C:\msdchem\1\DATA\DANOCSA3.D
Operator : CMS
Acquired : 9 Oct 2015 6:15   using AcqMethod STAND35.M
Instrument :   MSD 5975
Sample Name: DANOCSA3
Misc Info :
Vial Number: 25

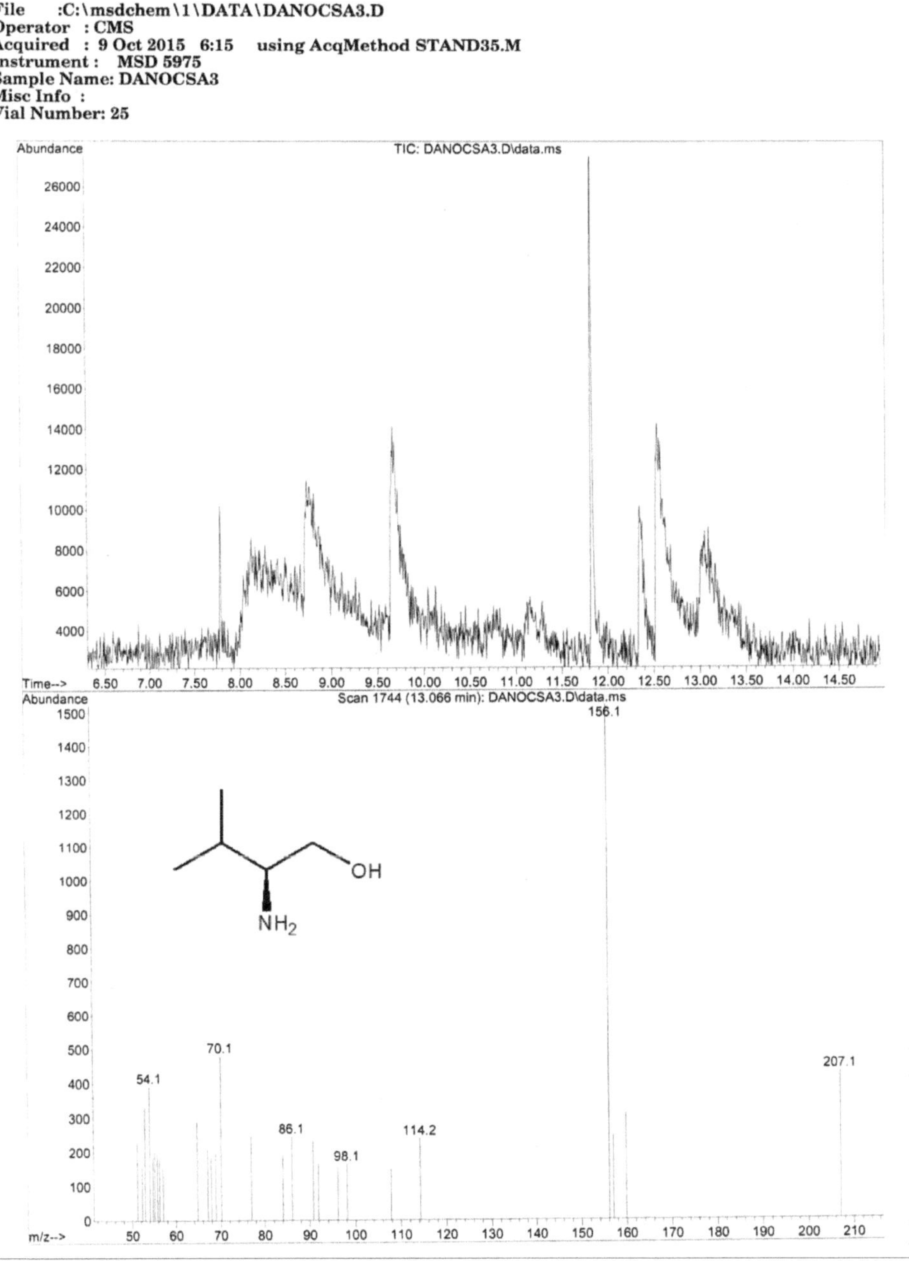

41

3.3.4 (*S*)-4-Isopropyl-2-oxazolidinone (**10**)[7]

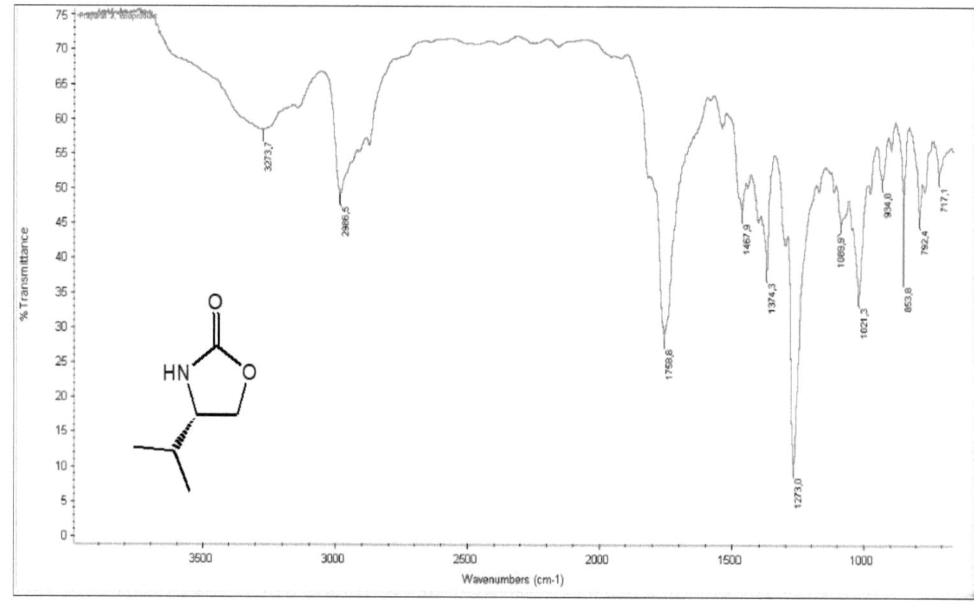

File   :C:\msdchem\1\DATA\DANOCSA4.D
Operator : CMS
Acquired : 8 Oct 2015 23:46   using AcqMethod STAND35.M
Instrument :   MSD 5975
Sample Name: DANOCSA4
Misc Info :
Vial Number: 42

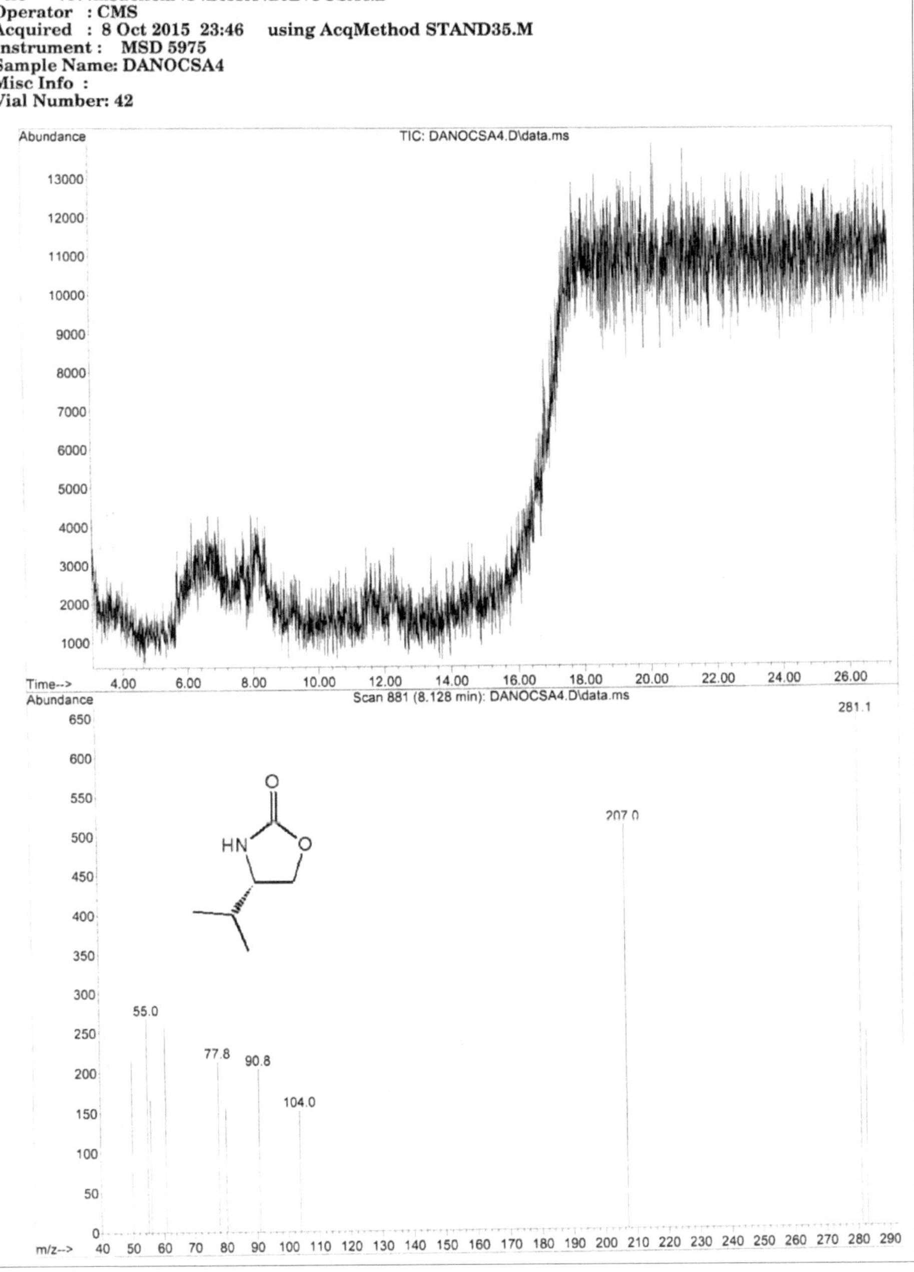

3.3.5 Di-*tert*-butyl-1-(2,6-diisopropylphenyl)-hydrazine-1,2-dicarboxylate (**11**)[9]

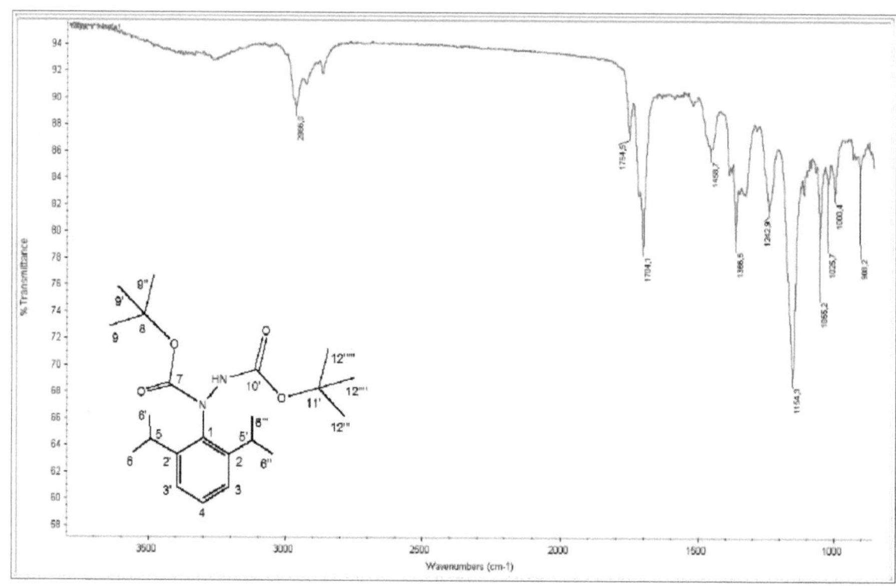

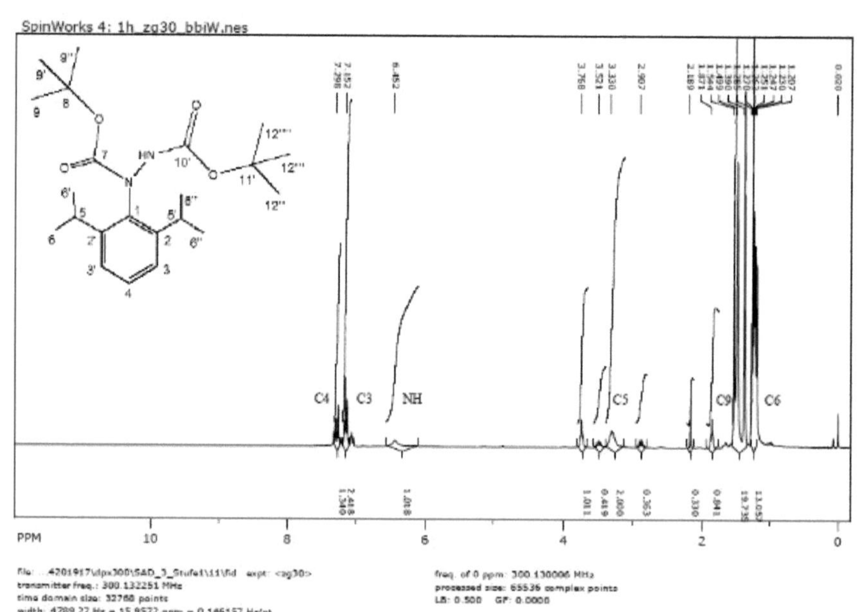

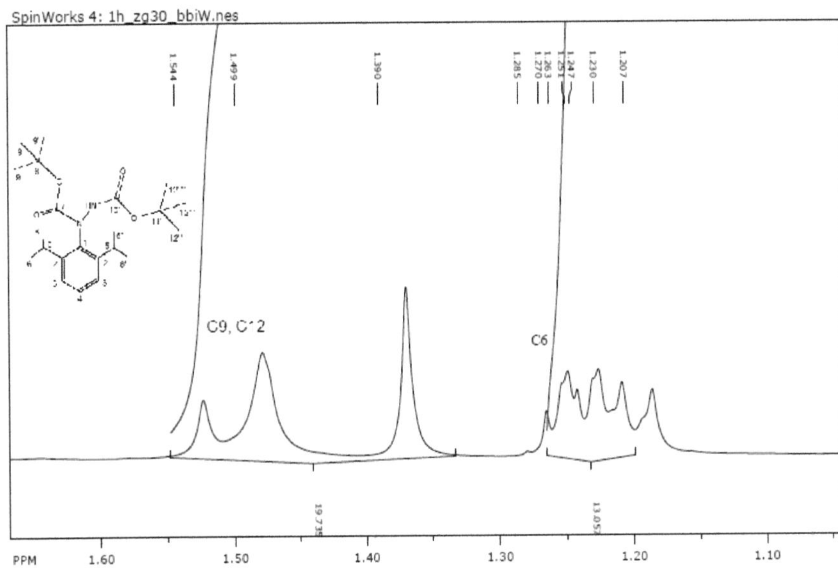

SpinWorks 4: 1h_zg30_bbiW.nes
1.544
1.499
1.390
1.285
1.270
1.263
1.251
1.247
1.230
1.207
C9, C12
C6
19.73
13.05
PPM   1.60   1.50   1.40   1.30   1.20   1.10
file: ...4201917\dpx300\SAD_3_Stufe1\11\fid   expt: <zg30>
transmitter freq.: 300.132251 MHz
time domain size: 32768 points
width: 4789.27 Hz = 15.9572 ppm = 0.146157 Hz/pt
freq. of 0 ppm: 300.130006 MHz
processed size: 65536 complex points
LB: 0.500   GF: 0.0000

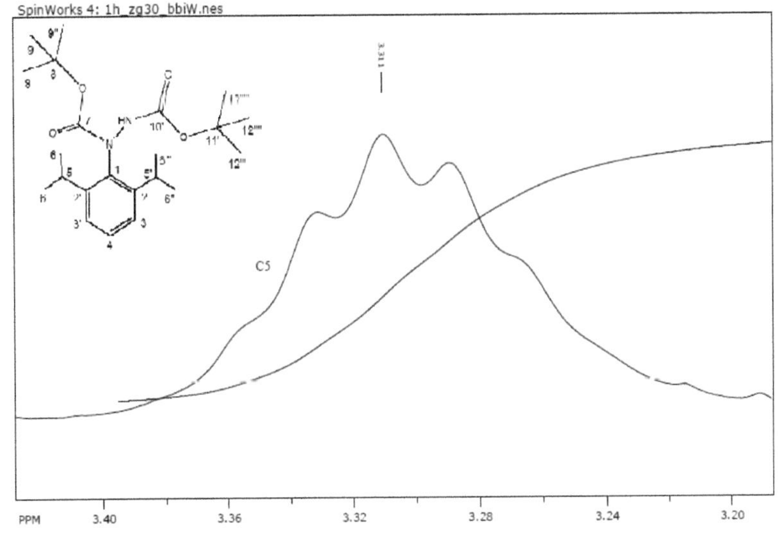

SpinWorks 4: 1h_zg30_bbiW.nes
3.311
C5
PPM   3.40   3.36   3.32   3.28   3.24   3.20
file: ...4201917\dpx300\SAD_3_Stufe1\11\fid   expt: <zg30>
transmitter freq.: 300.132251 MHz
time domain size: 32768 points
width: 4789.27 Hz = 15.9572 ppm = 0.146157 Hz/pt
number of scans: 32
freq. of 0 ppm: 300.130006 MHz
processed size: 65536 complex points
LB: 0.500   GF: 0.0000

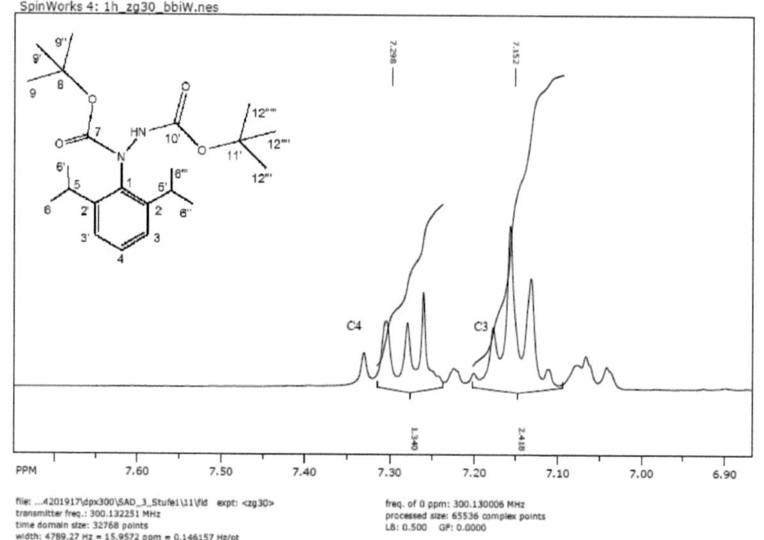

46

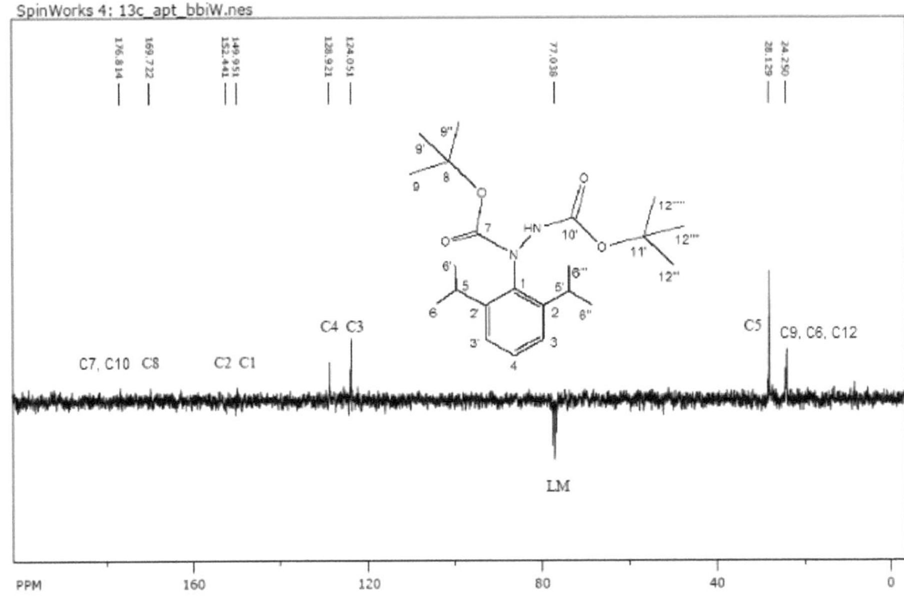

file: ...4201917\dpx300\SAD_3_Stufe1\10\fid   expt: <apt>
transmitter freq.: 75.476050 MHz
time domain size: 32768 points
width: 18115.94 Hz = 240.0224 ppm = 0.552855 Hz/pt
number of scans: 128

freq. of 0 ppm: 75.467749 MHz
processed size: 32768 complex points
LB: 3.000   GF: 0.0000

File   :C:\msdchem\1\DATA\dko\DKOsadik1.D
Operator : BEN
Acquired : 23 Sep 2015 15:17   using AcqMethod STAND50.M
Instrument :   MSD 5975
Sample Name: DKOsadik1
Misc Info : 1h60C
Vial Number: 11

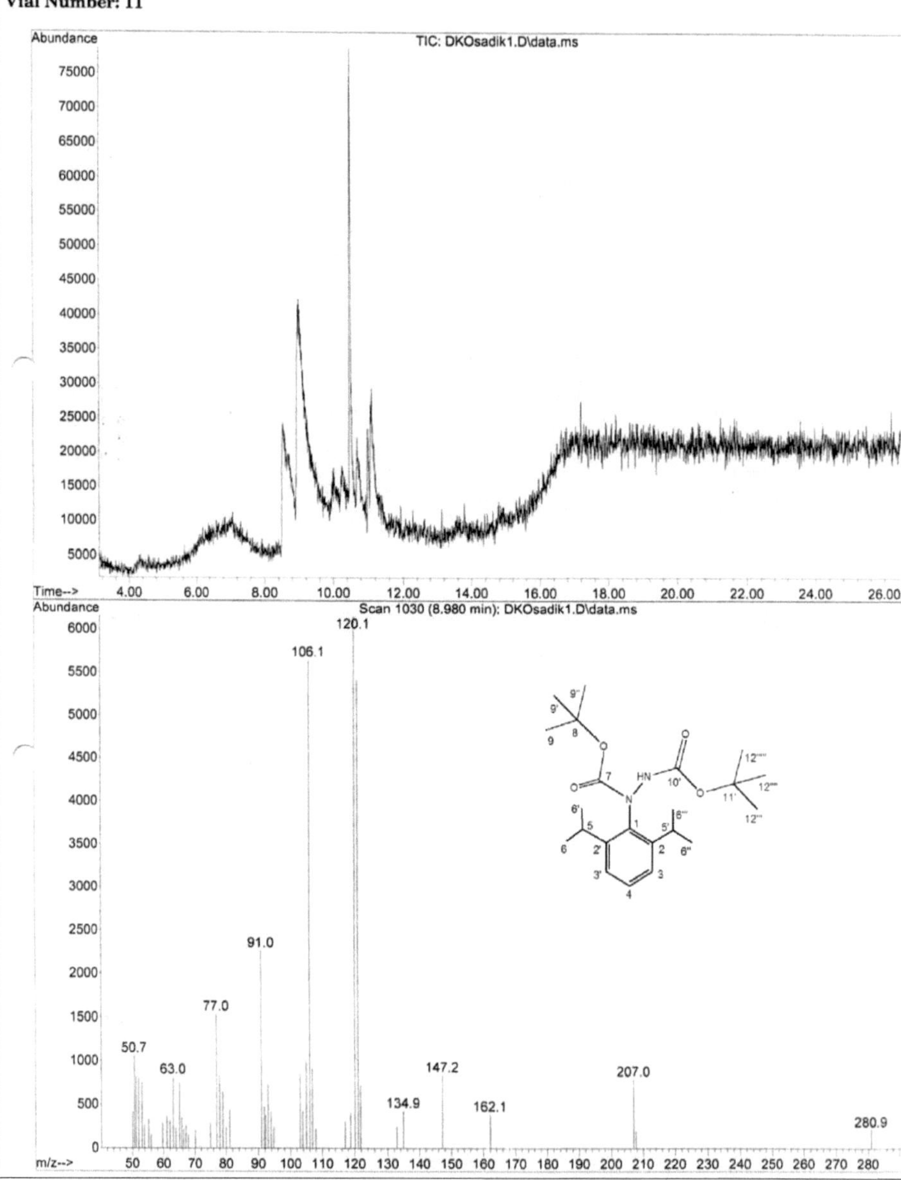

3.3.6 2-(2,6-Diisopropyl-phenyl)hydrazin-1-ium chloride (**12**)[9]

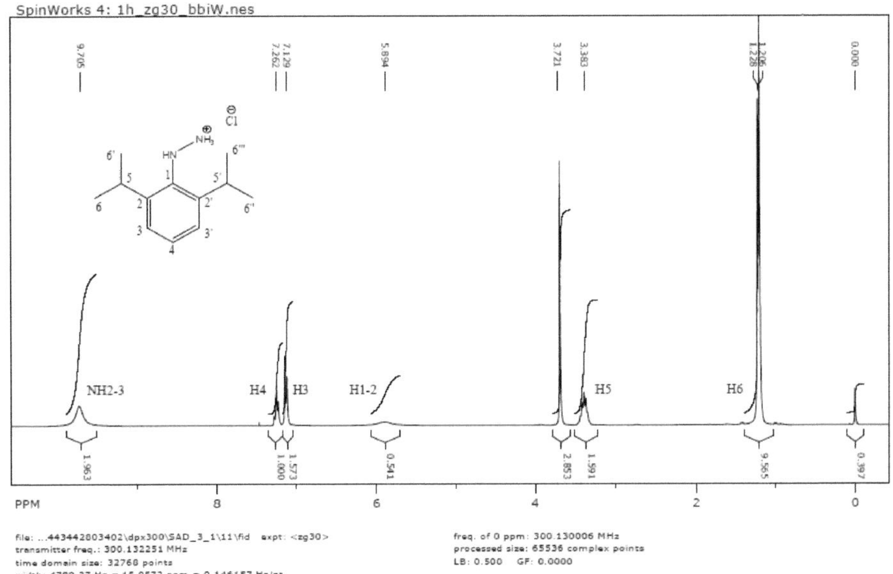

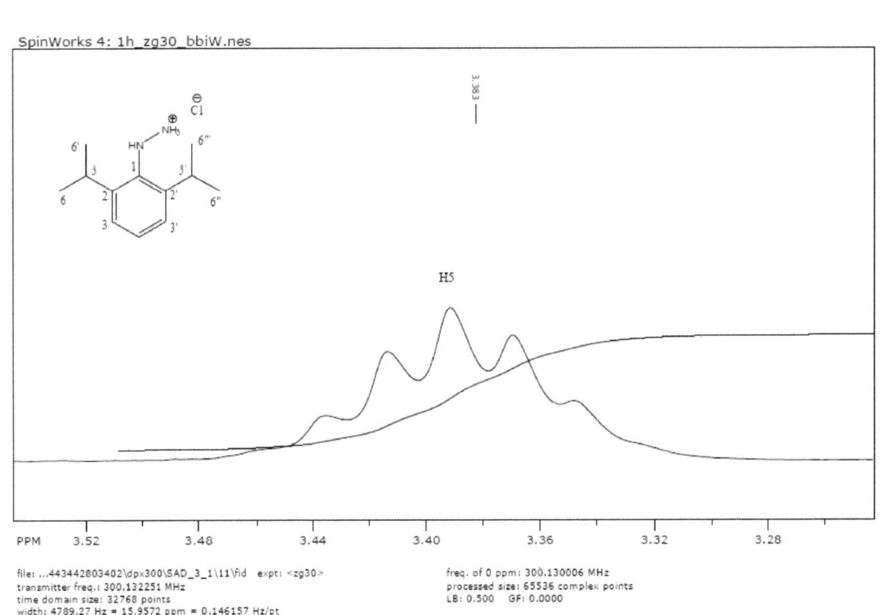

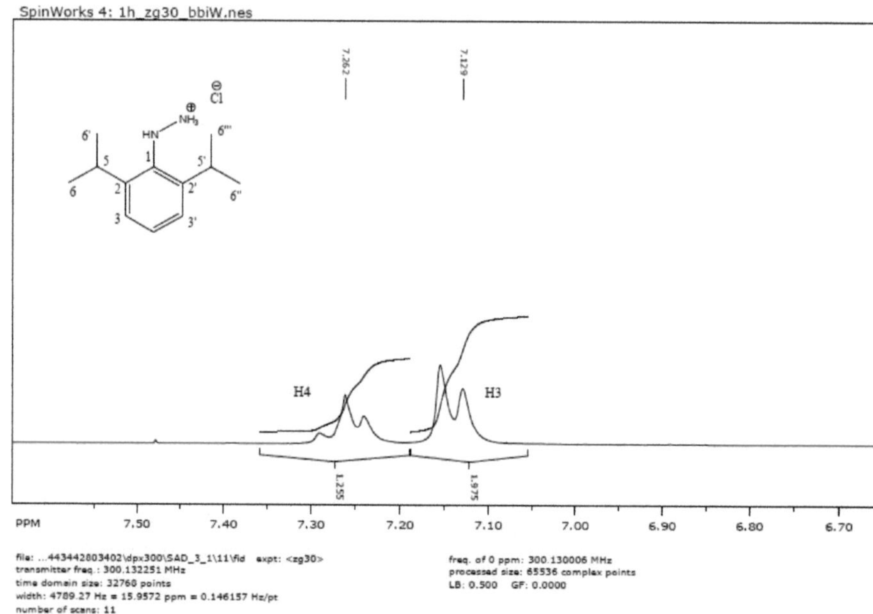

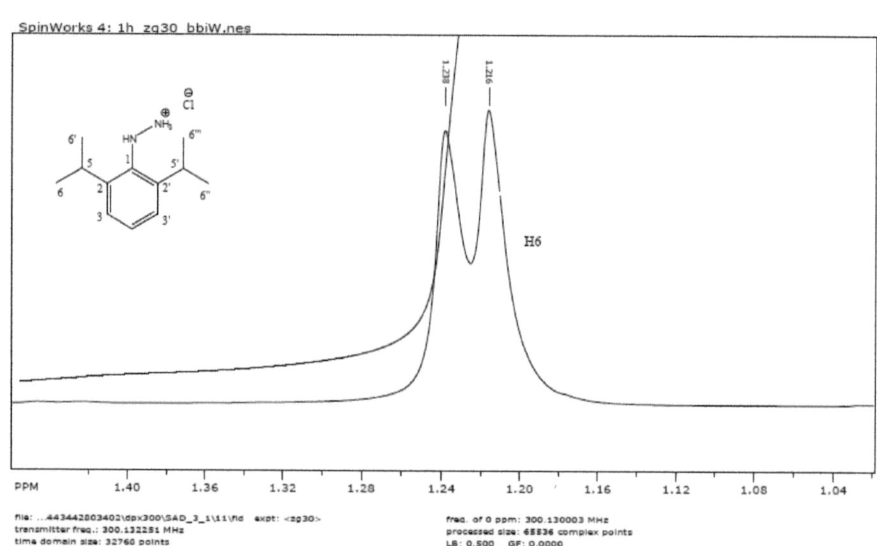

50

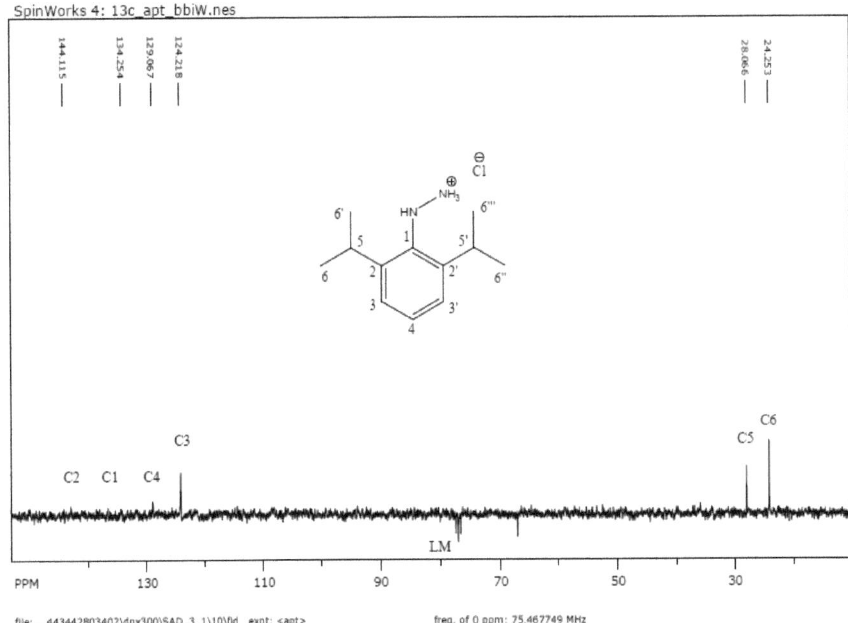

file: ...443442803402\dpx300\SAD_3_1\10\fid   expt: <apt>
transmitter freq.: 75.476050 MHz
time domain size: 32768 points
width: 18115.94 Hz = 240.0224 ppm = 0.552855 Hz/pt
number of scans: 128

freq. of 0 ppm: 75.467749 MHz
processed size: 32768 complex points
LB: 3.000   GF: 0.0000

File   :C:\msdchem\1\DATA\DANOCSA6.D
Operator : CMS
Acquired : 9 Oct 2015 5:04   using AcqMethod STAND35.M
Instrument : MSD 5975
Sample Name: DANOCSA6
Misc Info :
Vial Number: 23

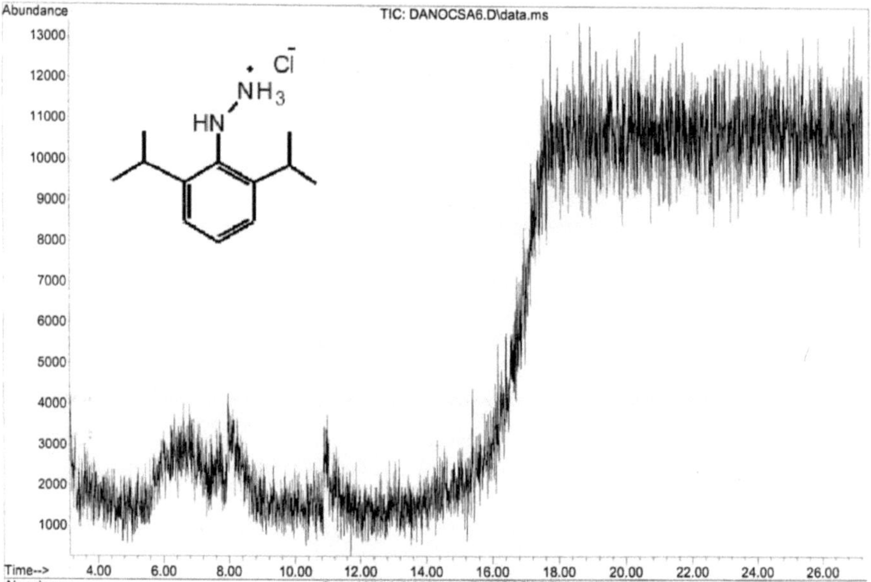

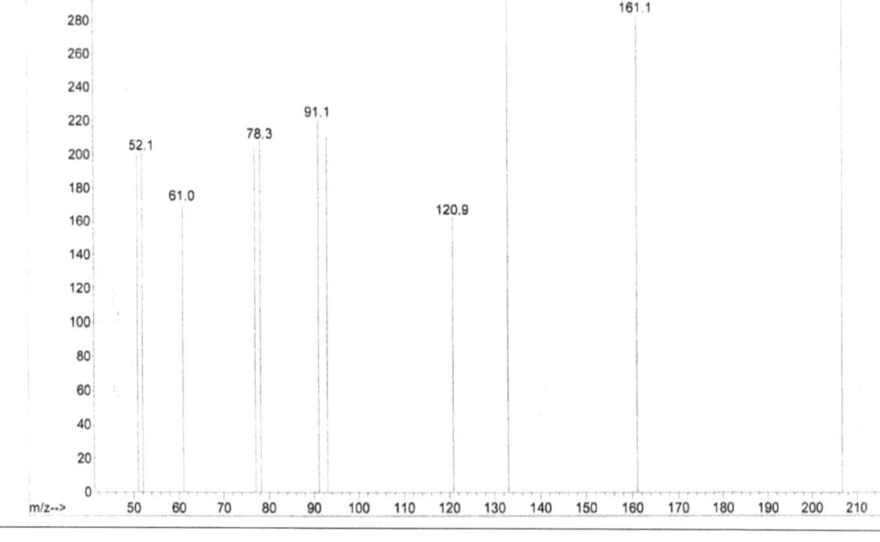

3.3.7 Dimethyl *o*-tolylboronate (**14**)[10]

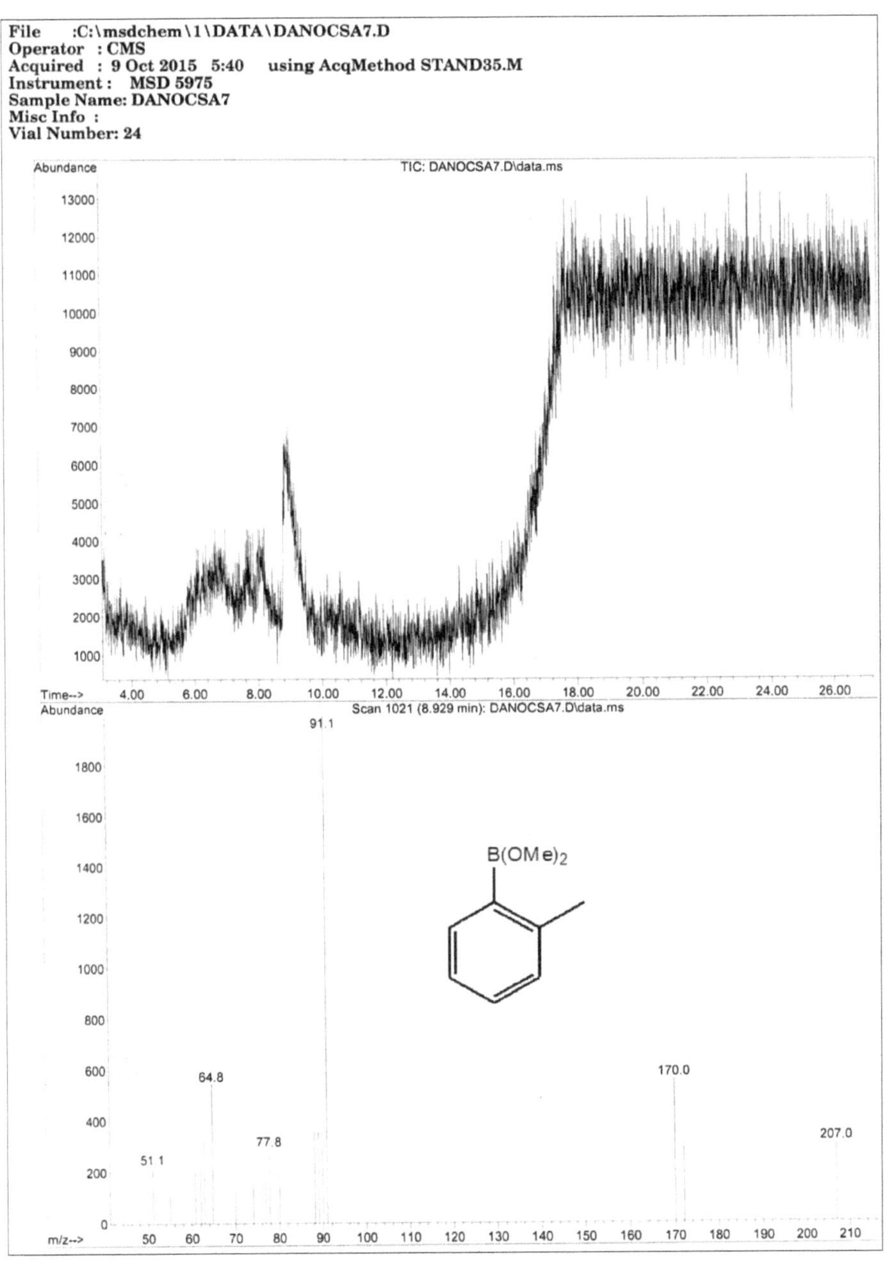

3.3.8 1-Methyl-4'-methoxybiphenyl (16)[11]

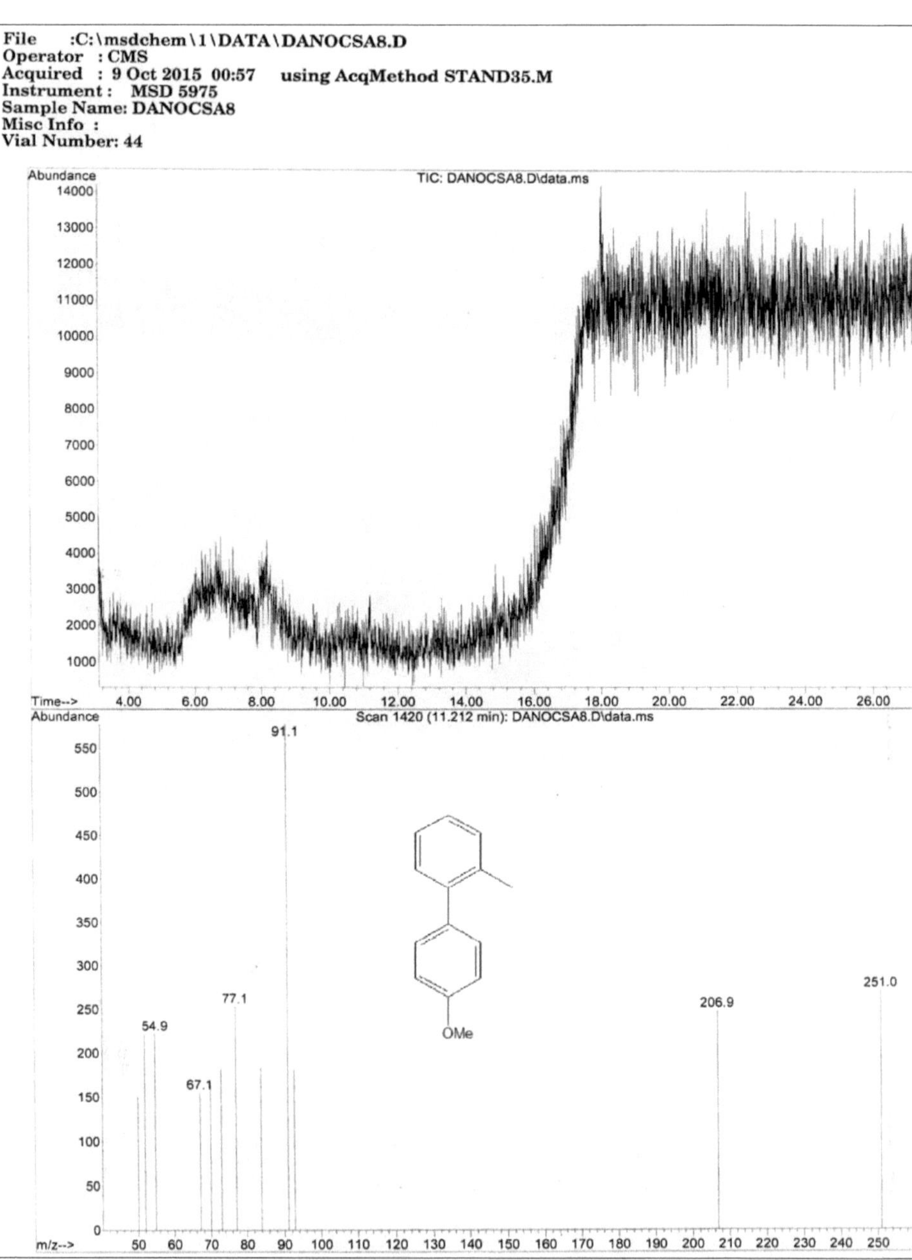

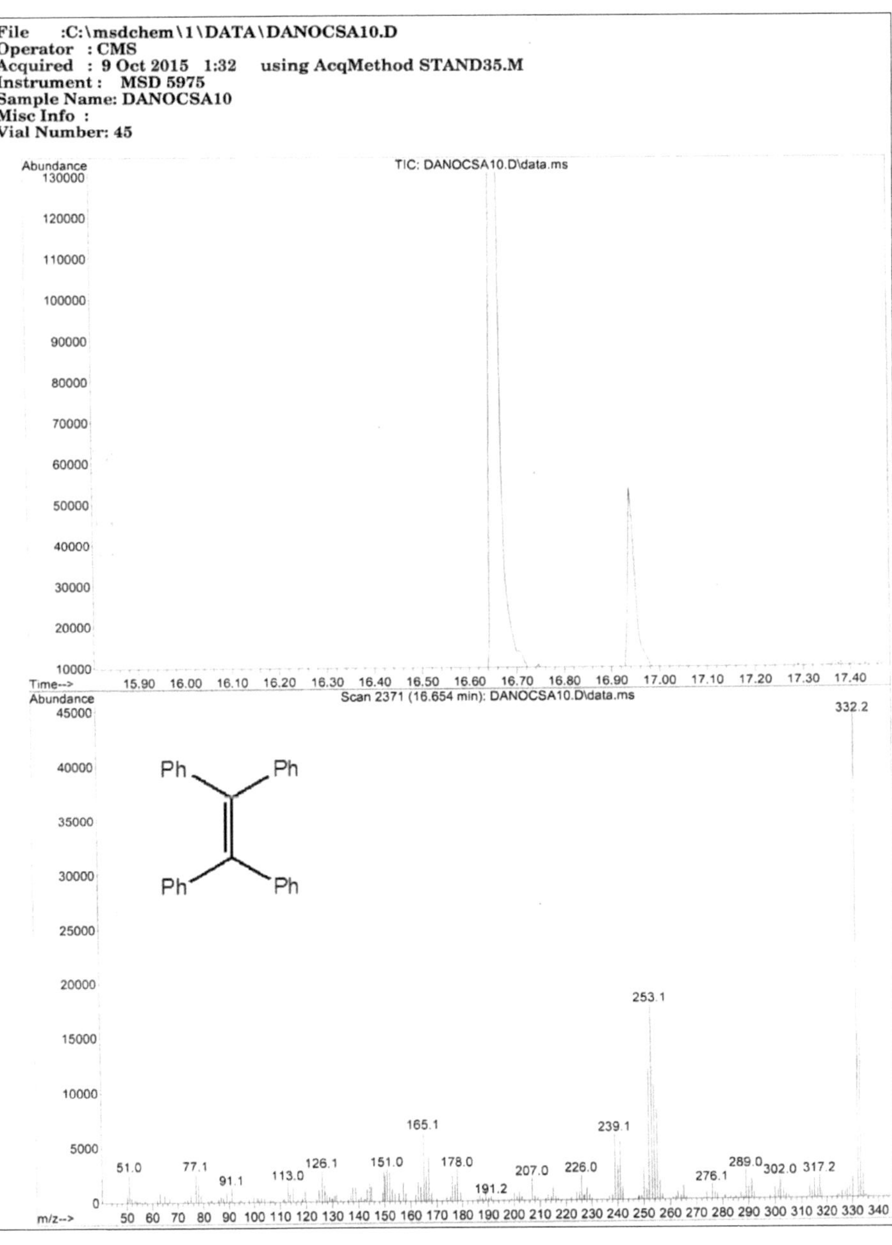

File   :C:\msdchem\1\DATA\DANOCSA10.D
Operator : CMS
Acquired : 9 Oct 2015 1:32   using AcqMethod STAND35.M
Instrument :   MSD 5975
Sample Name: DANOCSA10
Misc Info :
Vial Number: 45
Abundance
TIC: DANOCSA10.D\data.ms
130000
120000
110000
100000
90000
80000
70000
60000
50000
40000
30000
20000
10000
Time-->   15.90 16.00 16.10 16.20 16.30 16.40 16.50 16.60 16.70 16.80 16.90 17.00 17.10 17.20 17.30 17.40
Abundance
Scan 2371 (16.654 min): DANOCSA10.D\data.ms
45000
40000
35000
30000
25000
20000
15000
10000
5000
332.2
253.1
165.1
239.1
51.0
77.1
91.1
113.0
126.1
151.0
178.0
191.2
207.0
226.0
276.1
289.0
302.0
317.2
0
m/z-->   50 60 70 80 90 100 110 120 130 140 150 160 170 180 190 200 210 220 230 240 250 260 270 280 290 300 310 320 330 340
Ph
Ph
Ph
Ph

3.3.11 5-Hydroxy-3-oxo-5-phenylpentanoic acid methyl ester (23)[16]

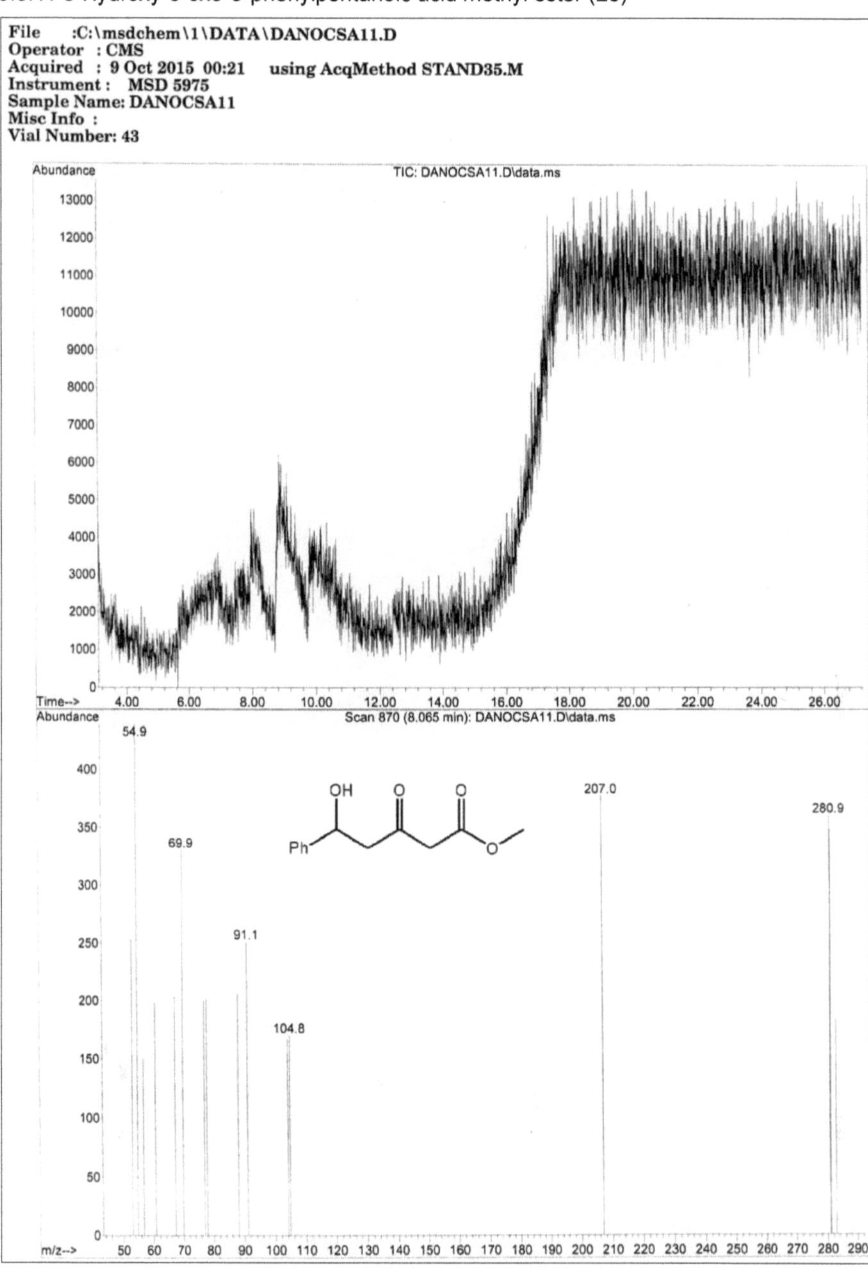

Zum Abschluss bedanke ich mich an dieser Stelle bei Herrn Prof. Dr. Ralf Giernoth für die Aufnahme in seinen freundlichen Arbeitskreis. Außerdem bedanke ich mich ganz herzlich bei den Assistenten Daniel von der Heiden und Jan Stein, die mich während der gesamten Praktikumszeit und darüber hinaus sehr gut betreut und unterstützt haben.

Außerdem bedanke ich mich bei Darius Kootz und Mathias Paul für die konstruktiven Gespräche und Ratschläge und für Ihre Unterstützung bei der Aufnahme und Auswertung der Massenspektren des untersuchten Materials.

Mein Dank gilt auch für Murat Atar für seine Hilfsbreitschaft bei der Aufnahme der IR-Spektren. Dankbar anerkennen will ich auch die freundliche und gute Zusammenarbeit meiner Mitpraktikanten sowie jede erdenkliche Hilfe von den Mitarbeitern und Mitarbeiterinnen des Instituts.